The International Library of Bioethics

Volume 115

The *International Library of Bioethics* comprises international and interdisciplinary volumes on foundational and applied issues in bioethics. The series focuses on perennial topics of aging, mental health, preventive medicine, medical research issues, end of life, biolaw, and other areas of bioethics, whilst expanding into other current and future questions involving the living world. The *International Library of Bioethics* examines the role of bioethics against the background of increasing globalization and interdependency of the world's cultures and government.

The series explores meaningful ways to answer the questions of how to direct biotechnology to human and other living things' ends, how to deal with changed values in the areas of religion, society, and culture, and how to formulate a new way of thinking. That is, to create a new, practical bioethics.

The international Library of Bioethics welcomes book proposals representing the broad interest of the series' focus. We especially encourage proposals addressing aspects of changes in biological and medical research and clinical health care, health policy, medical and biotechnology, and other applied ethical areas involving living things, with an emphasis on those interventions and alterations that force us to re-examine foundational issues.

Florencia Luna

The Ideal and Non-Ideal in Bioethics

Decision Making in an Imperfect World

Original Translator: Timothy Daly

Springer

Florencia Luna
Bioethics Program
FLACSO and CONICET
Buenos Aires, Capital Federal, Argentina

ISSN 2662-9186 ISSN 2662-9194 (electronic)
The International Library of Bioethics
ISBN 978-3-032-14593-2 ISBN 978-3-032-14594-9 (eBook)
https://doi.org/10.1007/978-3-032-14594-9

This work was supported by FLACSO / FACULTAD LATINOAMERICANA DE CIENCIAS SOCIALES (VAT 30-60267573-0).

Translation from the Spanish language edition: " Lo ideal y no ideal en bioética. Tomando decisiones en un mundo imperfecto" by Florencia Luna, © Eudeba 2025. Published by Eudeba. All Rights Reserved.

Translation of the work originally titled Lo ideal y no ideal. Tomando decisiones en un mundo imperfecto, published by Eudeba, Buenos Aires, 2025

This Springer imprint is published by the registered company Springer Nature Switzerland AG
The registered company address is: Gewerbestrasse 11, 6330 Cham, Switzerland

To my loving sons, Lucio and Marco, and to my dear niece Morena, in the hope that they may live in a fairer, more cooperative and closer to the ideal, world.

Preface

This book originates, on the one hand, from a prestigious lecture series organised by the Argentine Society for Philosophical Analysis (SADAF), with the aim of subsequently producing a publication. Unlike other centres or departments of philosophy worldwide that often regard applied philosophy with disdain, SADAF has always respected and incorporated bioethics as just another area of philosophy. I have always believed that bioethics should be nourished by both ethics and political philosophy, since together they provide the solid theoretical foundation necessary to conduct good bioethics—such as theories of justice, the analysis of exploitation, or ideal versus non-ideal theory. On the other hand, this book is the result of decades of work on some of these issues, such as the concept of vulnerability and the relationship between the ideal and the non-ideal.

Two people were fundamental to this English version. The first is Ruth Macklin, my much-loved colleague, friend and mentor. She is a person of rare humility and empathy, yet also possesses great inner strength and confidence. Undoubtedly, she has radiated the greatness of someone who had nothing to prove. She encouraged me to publish this book in English and offered invaluable suggestions, which I have tried to incorporate.

The second person I wish to thank in particular is Timothy Daly, who translated these pages and provided an excellent version, sparing the reader what would have been my own Tarzan-like attempt. He showed immense patience throughout the back and forth of our many drafts, suggested changes, and designed the Figure illustrating my decision-making framework. It has been a joy to work with him, benefitting from both his professionalism and generosity, as well as his wonderful attitude as part of the academic team at FLACSO.

And last but not least, I would like to thank the entire FLACSO team: its Director, Dr. Valentina Delich, whose support has been unwavering; the administrative team, especially Daniela Grebnicoff; and, of course, the dear and efficient Ana Gromick, who helps me continuously. I also wish to acknowledge those who read earlier versions and made insightful suggestions: Paola Buedo, Ignacio Mastroleo, and Mariana Cristina. Naturally, the final responsibility for what is presented here is mine alone.

Living in the Global South is often difficult. Things are not as they should be; we suffer countless problems that more affluent societies have already solved. The ideal but mostly the non-ideal pervade the everyday. Vulnerability is part of our daily lives. But living in this society also allows us to see the world from a different angle—from the periphery, far from the hegemonic view. Many of the ideas and proposals in this book could have a real impact on the lives of people who are not usually in the spotlight.

Finally, I must acknowledge the privilege of having a family that loves this South and my country Argentina, in which I grew up and was educated. So I would like to end by especially thanking my dearest sisters—my emotional bastions—Felicitas and María, my dear sons Lucio and Marco, and, although they are no longer with us, the idealist who was my father and the lioness who was my mother—two exceptional human beings.

Buenos Aires, Argentina Florencia Luna

Acknowledgement The research reported in this publication was supported by the Fogarty International Center of the National Institutes of Health under Award Number R25TW001605. The content is solely the responsibility of the author and does not necessarily represent the official views of the National Institutes of Health.

Competing Interests The author has no competing interests to declare that are relevant to the content of this manuscript.

I have tried, though I am not sure how successfully, to write straightforward stories. I dare not call them simple, for nowhere on Earth is there a sole page or even a word that is so, since each implies the universe, whose most obvious trait is complexity.
Jorge Luis Borges, Preface to Brodie´s Report (1979)

He intentado, no sé con qué fortuna, la redacción de cuentos directos. No me atrevo a afirmar que son sencillos; no hay en la tierra, una sola, página, una sola palabra, que lo sea, ya que todas postulan el universo, cuyo más notorio atributo es la complejidad.

Jorge Luis Borges, El informe de Brodie (1979) in Obras Completas (2011) Sudamericana, Buenos Aires.

Introduction

This book examines both enduring and emerging issues in bioethics. It offers a perspective informed by sustained engagement with, and contributions to, the field's evolution. Although these reflections have largely taken shape within the context of the Global South, it is my hope that they will resonate more broadly and contribute meaningfully to global bioethical debates.

Before engaging in a deeper study of ethical issues, I would like to point out some milestones that have made this book necessary. Bioethical thinking emerged in the 1970s, mainly in the Anglo-Saxon world, and its first concerns revolved around the clinic. Classical problems included the professional-patient relationship, euthanasia, and assisted reproduction, amongst others. Major technological advances—the ventilator, intensive therapies and widely available antibiotics—were accompanied by broader societal changes that raised the need to consider patients' points of view, broadening the scope of the field.

Since its beginnings, the type of approach used by bioethics has been relatively simple—Kantian or utilitarian theories seemed to exhaust the initial answers. One of the first classic texts by Beauchamp and Childress in 1979, *Principles of Biomedical Ethics*, proposed four *prima facie* principles that are still very popular. These principles became known as the 'Georgetown mantra' due to their near-religious status in the field and their automatic repetition. Beauchamp and Childress's proposal caused ethical analysis to be simplified and reduced to the application of some uncritical version of the principles of autonomy, beneficence, non-maleficence and justice. Some authors considered that bioethical theory could be exhausted by the mechanical application of these principles. In this nascent period, the permissibility of medical practices was questioned in the abstract. The kinds of problems were novel, and the debate between medicine and ethics was just beginning. In fact, a now classic article written by Stephen Toulmin (1986) argued that medicine "saved the life of ethics". Medicine would rescue ethics from irrelevant academic abstractions, eternal meta-ethical debates and endless controversies between duty-based ethics on one hand and different versions of consequentialism on the other.

Faced with these simpler and more naïve analyses of early bioethics, research ethics would grow in force towards the end of the 1990s. It was no longer merely informed consent that took precedence, but questions of justice and double standards that were being asked by a now burgeoning, international field. However, when it came to actually implementing research ethics, particularly when assessing research protocols, there was still a tendency for simplistic analyses and a reliance on checklists as a means of quality control. The approach went little further than checking whether informed consent had indeed been obtained, whether the studied population was vulnerable and other related questions. In other words, to do ethics was to simply tick the appropriate boxes.

Thus, research ethics, like early bioethics, has often avoided deeper analysis, relying instead on laws or guidelines. By doing so, it has endorsed what London has called procedural ethics (London, 2005, 2021). This meant neglecting substantive ethical analysis, now recognised as vital, including global justice and social determinants of health.

Although in some instances such checklists help or provide structure to ethical assessment, what has really been at stake is the depth of analysis, taking into account the particular situations and the specifics of each case at hand. It has become quite clear now that research ethics necessarily introduces more complex conceptual elements than previously existing analysis. To take just three examples, such ethical work has led to analyses of the levels of voluntariness of certain decisions in a research environment, more meaningful proposals of the notion of justice, as well as different analyses of the concept of exploitation. The international scope of research ethics demonstrates the need to incorporate theories of justice and political philosophy (Luna, 2005).

Another area that received very little early attention was public health ethics. Though it had pioneers of the stature of Ruth Faden in her foundational book (Faden and Powers, 2006), it was not until the COVID-19 pandemic that this type of approach found itself centre stage. Issues related to public health quickly showed the limits of the view based solely on individuals and their preferences, and a focus on the common good and collective welfare would emerge.

Thus, bioethics, its interests, and the possible answers to its questions have changed over a relatively short period of time. At the same time, the world is opening up to the present-day relevance of this new, evolving speciality. Little by little, the Anglo-Saxon viewpoint and the classical cookie-cutter approach are giving way to local perspectives and a host of different theories. Each new proposal for analysis—clinical, research, public health—introduces different types of theoretical elements that range from classical ethics to theories of justice and political philosophy. Theories of justice are adapted and applied to health systems in their variants of the welfare state (Daniels, 1988; Green, 1981) or of the neoliberal model (Engelhardt, 1995). There are more sophisticated approaches with respect to the notion of exploitation. To take just one notable example, Malmqvist's (2011) analysis of exploitation would come to criticise Wertheimer's (1996). This shows that reality is always imposing new demands on bioethics, which evolves as it rises to these new challenges.

In addition, in recent decades there has been much greater depth and sophistication of analysis, with approaches, such as feminism, that defy traditional analysis and require new theoretical tools. Different feminisms, in particular, have had a great impact on bioethics, generating highly attractive reflections and perspectives. Moreover, feminism has arguably enjoyed more respect within bioethics than in philosophy itself (Mackenzie, 2019). These views provide an ethical, as well as a political and critical, look at the *status quo*. But, in addition, they provide different perspectives on the same phenomenon since there are multiple variants, such as liberal feminism, egalitarian feminism or radical feminism, with each approach emphasising different elements. In tune with these new developments, this book addresses the current richness of this fascinating discipline and delves into its very foundations.

Keeping in mind all of these developments, the aim of this book is to move away from the easy and simplifying version of bioethics. For example, it avoids 'theory of principles' analyses and checklists, and moves on to more complex reflections. It attempts to grapple with the most cutting-edge proposals in political philosophy and ethics, including recent approaches that discuss the very ideal (or not, as the case may be) of ethics and political philosophy.

Indeed, just as centuries of ethical reflection led to the development of meta-ethics, I now consider that the time is ripe for addressing the theory on which bioethics is based, i.e., it is time for meta-bioethics. This involves theoretically analysing how bioethics should, and does in fact, proceed. For example, it involves examining how ideal its reflections should be and how much account ought to be taken of reality and its obstacles. We must deliberate about whether we necessarily have to choose only one field (the ideal or the non-ideal) or, rather than falling into that typical dilemma, establish criteria to decide in which circumstances an answer can be given from the non-ideal, or from the ideal. This book tries to depart from the trodden path of dichotomies and think, instead, in terms of the complexity of the real world. This, in a way, implies continuing with the critical spirit of feminism. It means that I will take great pains to avoid intellectual straightjackets and will instead attempt to provide richer, more complex and dynamic answers to the problems the field faces.

Yet, why the need for such a theoretical exercise? It is not put forward as intellectual musing or philosophising for its own sake. It is instead based on Kurt Lewin's maxim that "there is nothing as practical as a good theory" (Lewin, 1943).[1] For instance, what we take to be the right relationship between the ideal and the non-ideal will bear strongly on the kind of solutions we propose to concrete examples. How we conceptualise vulnerability will influence how we approach supposedly vulnerable populations and why we might refer to types of vulnerabilities rather than populations. If my hypothesis is correct, then such a theoretical analysis is both relevant and necessary if we are to claim that bioethics has reached the status of a mature discipline. This is why we must begin to take a critical look at the theoretical development of bioethics. These are issues for which analysis and reflection are well overdue.

This book has three main parts. Part I, "The Ideal and Non-Ideal: Theory and Practice", has two chapters that deal with the theoretical approach to the ideal and non-ideal and sets the ground for the rest of the analysis.

---

[1] I want to thank Timothy Daly for this excellent quote.

Within these pages I am wagering on the benefits of the arduous task of assessing when to appeal to idealised models and when to deal with the second best or with non-ideal proposals. There is often a tendency to assume there is only one true option and discard the other. Those who defend the virtues of the ideal point out the importance of having clear paradigms, that ideals should govern us, and look down on muddy everyday existence and its embarrassment of barriers that hinder the path to the ideal. But those who are in the trenches of reality, those who must deal with problems embedded in particular socioeconomic situations, may feel that such ideal proposals are impossible to carry out. They face a road full of obstacles that they cannot ignore; those noble and longed-for ideals seem totally unattainable! But by criticising ivory tower idealists, they themselves get stuck in the opposite trench and argue that there is no place for any model or ideal. Yet sometimes it really is necessary to go beyond the myopia of the *status quo* to look for deeper changes, that is—to appeal to the ideal. And we need a clear northern point on our compass. One of the big problems, in my opinion, is that decision making cannot be arbitrary, since our actions and decisions necessarily have, or at least, aspire to, a certain coherence. Obviously, it is more difficult to make these judgements, but I stake my claim on arguing that this is what should be done: taking what is valuable and relevant from each position.

The first chapter proposes the exercise of analysing two cases using a decision-making mechanism, with the aim of illustrating the challenges of the real world and the paradoxical need to resort to both ideal and non-ideal responses. As the decision-making mechanism and the cases unfold, the chapter gradually introduces clues, assumptions and theoretical considerations that are explicitly developed in the following chapter.

The first case exemplifies non-ideal decision making. It is a case of conscientious objection following a request for a legal abortion after rape. In such cases, a non-ideal theory must be used. Legal abortions in cases of rape in Argentina that were denied before the enactment of the current law are examined.[2] In those situations it was impossible to apply the ideal because those who objected did so in such a way as to prevent the provision of safe and legal abortion in a timely manner. Objectors had so much power that pregnant women or pregnant persons, more often than not, were disrespected.

The second case deals with the situation of research with pregnant or potentially pregnant people and the absence of legal and safe abortions in case they have to resort to them. This is a situation that concerns many developing countries and which is becoming increasingly relevant in the USA with the reversal of the Roe *v.* Wade ruling in June 2022. In these situations, on the contrary, one must appeal to the ideal theory. Ideality is represented by the high standards of the CIOMS (2016) *Ethical Guidelines* and their requirements when conducting research with people who may become pregnant.

---

[2] I first made the case for this approach in 2014. It was precisely because of the non-respect for reproductive rights that I began to address the problem of non-ideality (Luna, 2015).

These two examples present a number of points for consideration that are further explored in the next chapter. First, how reality forces us into more complex reflections. Second, how we cannot stay in any one single trench. Third, how the decision mechanism presented guides the choice of the right answer. Fourth, that this mechanism and the varied answers it provides assumes theoretical positions that should be made explicit.

Chapter 2 is the book's most theoretical chapter. It has a meta-bioethical focus, and considers ideal and non-ideal theories in detail. Here, the debates of political philosophy are adapted to bioethics in a series of reflections on the ideal and the non-ideal that will form the theoretical backbone of the whole book. The chapter reviews the different conceptual distinctions and some of the debates and criticisms of the ideal and the non-ideal. If we assume there is a multiplicity of theories with a different degree of ideality that are valid and that there are valuable elements we can rescue from the ideal and non-ideal, we make sense of the need for a decision-making mechanism such as the one proposed as a first step. I am treading new and exploratory ground here, as to my knowledge, this type of analysis has not been developed in bioethics—even less so with an inclusive vision of both proposals.

Part II, "Global Justice in a Non-Ideal World", focuses on the non-ideal world and its response to the call for global vaccine allocation during the COVID-19 pandemic. These three chapters examine the different responses to the global distribution of medical supplies in light of the debate between ideal and non-ideal theory, the decision-making process and the various theoretical proposals (such as dynamic duties or obligations) presented in Chapter 2. The analysis is approached with the objective of avoiding future blunders by learning from the mistakes made.

Chapter 3 analyses the context in which the pandemic occurred. It describes the Hobbesian state of nature in which countries were engaged in a kind of every-man-for-himself power struggle. The possibility of cooperation or altruism seemed completely unrealistic. This climate created a prisoner's dilemma where selfish decisions undermine collective benefit. However, even if one can read the lived reality as a prisoner's dilemma, the empirical situation shows that it could have been avoided if selfish and short-term power-grabbing had in fact not prevailed. To understand the behaviour of the different countries and their governments, the pressures and expectations of individuals and residents along with the role of different ideologies—whether nationalistic or cosmopolitan—are considered. The cooperative, supportive and global approach of the cosmopolitan vision brings us closer to the ideal. However, there are obstacles to reaching this ideal and unfortunately, this option does not seem viable. Therefore, this chapter deals with the second-best option amongst the non-ideal. It shows that if we apply our decision-making mechanism, even a model based on more moderate forms of nationalism suffers from serious shortcomings.

Chapter 4 goes on to explore how to solve the equity dilemma of global vaccine distribution. It analyses which initiatives would make the response more ethical and fairer. To do this, I examine the proposal for bilateral donations and then study in relative detail the COVAX platform proposed by GAVI (Global Vaccine Alliance), CEPI (Coalition for Epidemic Preparedness Innovation) and WHO (World Health Organization) together with other stakeholders. Here, not only the empirical status of both proposals is presented, but also their theoretical implications, for example, the assumptions that the different distribution models include and their impact on the effective delivery of vaccines. To this end, both the proportional distribution model, based on country population size, and the fair priority model, based mainly on the need and the level of severity of the pandemic in different nations, are reviewed. Again, theory is intimately tied up with practice: whether we prioritise population size or the needs of a country will alter the order of vaccine distribution. Although I consider that the COVAX mechanism undoubtedly implied an advance over the Hobessian struggle previously put forward, I also highlight the many shortcomings and problems faced by this proposal. The possibility of an improved COVAX or, as I call it, COVAX Plus, model is suggested.

Chapter 5, the last in this Part, revisits another of the theoretical notions I present in Chapter 2: dynamic obligations. It incorporates a more ideal view of the problem. It tries to show how new proposals (dynamic obligations) could open the way to less non-ideal answers. After applying the decision-making mechanism, presenting different responses provided in practice and also pointing out their many shortcomings, this chapter deals with a proposal that could not be fully implemented: a patent waiver policy. Although an attempt was made to implement it, the response came too late. This chapter examines its implications. Continuing with the suggestion to consider dynamic obligations, I present the draft versions of the international agreement signed by the WHO. The many theoretical, practical and political challenges facing this new instrument are reviewed. Dynamic obligations show how the ideal must inform assessments and demands for change. We need not accept and remain with the *status quo.* And both the provisional waiving of patents and a strong and equitable international agreement are novel elements that can help change reality.

The third and last part, "Other Non-Ideal Conceptions", is made up of three chapters and is dedicated to the concept of vulnerability and to a particular conception of autonomy. The analysis of vulnerability and relational autonomy, also introduced in Chapter 2, necessarily leads us to the realm of the non-ideal. The paradise of the ideal remains distant. Thus, these chapters are devoted to showing another aspect of the non-ideal.

Chapter 6, the first in this Part, analyses the traditional conception of vulnerability (criticising, among other things, its rigid analysis and assumptions) and proposes the metaphor of layers of vulnerability. The latter can be read with non-ideal lenses. It considers the context and the challenges it imposes. This chapter takes up the analysis I formulated in 2008 but complements it with the latest developments I published a decade later (Luna, 2009, 2019). It incorporates the analysis of vulnerability as a disposition and highlights the relevance of the stimulus condition that can 'trigger' a layer of vulnerability. Again, it will show how these theoretical perspectives have practical implications. It also considers the existence of layers of vulnerability that have a cascading effect and can generate incremental damage. This chapter also presents a key concept from feminism, intersectionality—another proposal closely connected with the non-ideal—and compares intersectionality and vulnerability.

Chapter 7 focuses on the practical consequences of the theoretical analysis presented in the previous chapter. It is yet another example of the importance of theoretical analysis for practice. One of the major criticisms of my initial analysis in terms of layers is taken up here: that decision makers were left with a multiplicity of vulnerabilities and few tools to assess and prioritise them. In this chapter I explain how vulnerability can be assessed and which elements should receive special attention. In addition to the layer metaphor, I develop the practical implications of the theoretical concepts of stimulus condition and vulnerability with cascading effects. They help us to prioritise and judge which layer of vulnerability to start with or which should be eradicated or minimised.

Finally, the eighth and last chapter of this book deals with another notion that is highly relevant in feminist ethics. It considers the concept of relational autonomy, which is closely connected to the non-ideal. While traditional analyses have often formulated a counterpoint between vulnerability and autonomy, this chapter shows how they can complement each other and provide more complete and interesting analyses for medical practice. It begins by articulating the problems and idealisations of the traditional concept of autonomy and the different answers provided by feminism. It also presents the tension that arises with Martha Fineman's approach, which gives great importance to the concept of vulnerability to the detriment of autonomy, and the 'rescue' and valorisation proposed by Catriona Mackenzie. Autonomy, even with its challenges, is undoubtedly not only an interesting concept but also a crucial one for women's self-determination. Then, I outline approaches such as those of Susan Sherwin, Mackenzie or Marina Oshana and point out the different levels of demands that feminist proposals may imply and how the context has a role. Finally, this first part of the chapter closes with proposals coming from the periphery or from the South. The last section addresses the problem of autonomy and vulnerability but in relation to a practical situation, such as people with mental illness, and shows how a relational vision of autonomy can complement an analysis of vulnerability based on layers, and provide extremely interesting answers in practice.

The book ends with an Epilogue where I reconstruct the main arguments provided in these pages and show how bioethics has to be aware of the ideal and non-ideal theories, concepts and answers it provides.

Thus, I hope that this intellectual journey through some key approaches and notions of bioethics will show the importance of good theoretical analyses and the contributions they can make to this discipline. These considerations are not only directly relevant to those who find themselves in the trenches of medical decision making, but also for those of us on the other end who wish to be treated ethically and with respect.

References

Beauchamp, T. and Childress, J. (1979) *Principles of Biomedical Ethics*. New York. Oxford University Press. (Translation of the fourth edition (1999) *Principios de ética biomédica*. Barcelona, Masson, S.A.)

Council for International Organizations of Medical Sciences (CIOMS). (2002) (2016). *International Ethical Guidelines for Health-Related Research Involving Human Subjects* (3rd ed.) Geneva: CIOMS.

Daniels, N. (1988). *Just Health Care*. Cambridge. Cambridge University Press.

Engelhardt, H.T. (1986). *The Foundations of Bioethics*. Oxford University Press

Faden, R.R. and Powers, M. (2006) *Social Justice: The Moral Foundations of Public Health and Health Policy*. Oxford University Press.

Green, R. (1981). "Health Care and-Justice in Contract Theory Perspective." In Shelp, Earl (ed.), *Justice and Health Care*, Boston.

Lewin, K. (1943). "Psychology and the process of group living", *Journal of Social Psychology*, 17, 113–131.

London, A. J. (2005). "Justice and the Human Development approach to International Research", *Hastings Center Report*, 35 (1) 24–37.

London, A. J. (2021). *For the Common Good*. Oxford University Press.

Luna, F. (2005). "Poverty and Inequality: Challenges for the IAB", International Association of Bioethics Presidential Address, *Bioethics*, vol. 19, No. 5/6, pp. 431–459.

Luna, F. (2009). "Elucidating the Concept of Vulnerability. Layers not Labels", *International Journal of Feminist Approaches of Bioethics*, vol 2, No. 1, spring.

Luna, F. (2015). "Medical ethics and more: ideal theories, non-ideal theories, and conscientious objection," *Journal of Medical Ethics*, vol. 41, pp. 129–133.

Luna, F. (2019). "Identifying and evaluating layers of vulnerability—a way forward", *Developing World Bioethics*, 19: 86–95. Online: https://onlinelibrary.wiley.com/doi/epdf/10.1111/dewb.12206.

Mackenzie, C. (2019). "Feminist innovation in philosophy: Relational autonomy and social justice", *Women Studies International Forum* 72 (144–151).

Malmqvist, E. (2011). "Understanding exploitation", *IRB: Ethics & Human Research 33* (2) 1–5.

Toulmin, S. (1982). "How medicine saved the life of ethics", in *Perspectives in Biology and Medicine*, Summer; 25(4):736–50.

Wertheimer, A. (1996). *Exploitation*. Princeton University Press.

Contents

Part I
The Ideal and Non-Ideal: Theory and Practice

Chapter 1
Contradictory Responses to Similar Cases?

Abstract The first chapter proposes the exercise of analysing two cases about reproductive rights using a decision-making mechanism with five criteria. The aim is to illustrate the challenges of the real world and the paradoxical need to resort to both ideal and non-ideal responses. The issue is why we must resort to such different kinds of theories, whether this is contradictory or problematic, and to assess how justified it is to draw on theories of such differing natures The first case exemplifies non-ideal decision making. It is a case of conscientious objection following a request for a legal abortion after rape. In such cases it is argued that a non-ideal theory must be used. The second case deals with the situation of research with pregnant or potentially pregnant people and the absence of legal and safe abortions in case they have to resort to them. In the latter situations, on the contrary, one must appeal to the ideal theory.

Keywords Ideal and non-ideal theories · Reproductive rights · Conscientious objection · Research with pregnant individuals · CIOMS (2016) Ethical Guidelines · Decision-making mechanism

1.1 Introduction

Our world is a complex one, and when we are faced with the task of resolving cases, shaping public policy or offering solutions to the various problems we encounter, we do so with different degrees of idealisation. In the following pages, I invite you to examine two cases that demonstrate challenges in practice with a strong theoretical counterpart, while offering some clues or clarifications along the way.

The two cases discussed here are addressed in very different ways: the first is approached using a non-ideal theory (NIT), while the second relies on an ideal theory

Based on my article, Luna, "Ideal and non-ideal theories: the challenges of decision-making in an imperfect world", 2021 in Victor, E. and L. Guidry-Grimes (eds), *Applying non-ideal theory to bioethics: Living and dying in a nonideal world*. Springer: Philosophy & Medicine Series No. 139. pp.17–40.

F. Luna, *The Ideal and Non-Ideal in Bioethics*, The International Library of Bioethics 115, https://doi.org/10.1007/978-3-032-14594-9_1

(IT). This book considers why we must resort to such different kinds of theories, whether this is contradictory or problematic, and assesses how justified it is to draw on theories of such differing natures—so that we may later uncover the assumptions involved. The present chapter does not, however, provide a detailed explanation of what an ideal theory is, nor does it lay out the various theoretical debates it entails; it also refrains from going into too much theoretical depth (that will be the task of the next chapter). So, readers who wish to investigate and explore these issues further can either turn to the next chapter now and return to this one later—or they can hold their breath and wait.

The focus of this chapter is on reproductive rights. The first case examines the issue of conscientious objection. In my 2015 article in which I dealt with conscientious objection in situations of legal abortions in Argentina (Luna 2015), I defended the need to use NITs and I formulated four requirements for selecting the best NIT from the three presented in that article. Here, the non-ideal is compared to the ideal proposal, which demonstrates the continuing relevance of the non-ideal response.

The second case considers the conduct of first clinical investigations with pregnant women in countries where abortion for foetal malformation, or what is sometimes called 'eugenic abortion', is not legally permitted and/or not accessible. Unlike the first case, it is argued that this situation must be solved with a more idealised IT or proposal. Both cases exemplify the practical value of combining NIT and IT to address real-world issues. Conversely, discarding either domain (ideal or non-ideal) can limit analytical possibilities. However, is this kind of analysis justifiable? Can we work with both kinds of theories? What does this position suppose? In this book we explore answers to these challenges. In what follows I also explain a decision-making mechanism to address concrete or specific ethical issues.

1.2 Ideal Theories and Bioethics: Very Brief Clarifications

As bioethicists, we ask ourselves, first, whether we should work with an IT or NIT. For example, should I spend my time thinking about what would constitute a fair health care system on the assumption that we will have full compliance? Or should I rather spend my time thinking about what every doctor should do against the backdrop of an unjust system? That is, do I analyse the failures of institutions and what they should be like to live up to IT, or on the contrary, start from the basis of everyday failures and non-compliance and, from there, analyse what would be the right thing to do in that particular and specific situation? Second, when a specific ethical problem arises, should I resort to ideal or non-ideal theory or idealised policies as a guide?[1] The first point is beyond the scope of this chapter, so I will work on the second problem.

Another important issue to clarify is the type of conceptual element we are working with. Institutional policies, regulations or even the resolution of cases, may exemplify more or less ideal theories. This means that in bioethics we do not usually deal with

[1] I thank Joe Millum for this comment.

'pure' ITs.[2] However, case resolution, policy making, etc. may involve idealised ITs or theories. An absolutely pure model rarely exists in concrete situations because of imperfect background elements. Thus, this chapter considers that we are dealing with an ideal theory in bioethics when those guidelines, regulations or public policies follow rules or principles that would be required in an ideal world, even if it falls short of being a pure IT. Conversely, an NIT in bioethics adjusts those rules to the limits of the real world and seriously considers feasibility limitations. In what follows, then, I assess which theory (NIT, IT or more or less idealised theories) should be applied to concrete examples involving reproductive rights in non-ideal contexts.

The final terminological clarification has profound conceptual connotations. Throughout the book I use the most neutral language possible with an inclusive vision; however, this chapter discusses cases that refer to the lack of respect for women's sexual and reproductive rights. Although the cases also include people with other gender identities with gestational capacity, pregnant women as well as women, the backdrop, especially in the discussion on abortion (Luna 2018), is that of discrimination and disrespect specifically towards women and particularly directed towards women of limited resources who face targeted discrimination (Ariza and Saldivia 2015). By contrast, the second, or research case, can be more inclusive of people with diverse gender identities. Therefore, while taking a broad and inclusive view of the issue, I occasionally refer to women, as I believe it is important not to lose sight of this discriminatory aspect of anti-choice regulations.

1.3 A Decision-Making Mechanism to Navigate Ideality

This section describes a decision-making mechanism that may serve a heuristic function and bring greater consistency to our work in bioethics. It also represents a first step in showing that working with two such different types of theories is not, in fact, contradictory. As far as I am aware, nothing of the sort has previously been developed in the field of bioethics.

In 2015, when I was working on an article on conscientious objection (Luna 2015), I was faced with a situation that could not be adequately resolved through IT or the most idealised proposal, and I had to evaluate how to decide between different NITs. At that time, four criteria or requirements were proposed, taken from a reconstruction formulated by John Simmons and John Rawls (Simmons 2010: 18–19; Rawls 1999, Sect. 2: 39, 69), to compare and select an NIT from among several candidates (Luna 2015: 132).[3] These criteria can also be extended to the comparison between NITs and ITs. These requirements are:

R1: The proposal should be politically possible;

[2] For example, Robeyns considers the design of actions and the implementation of measures as part of non-ideal work (2008: 343).

[3] In this version I will put them in a different order as I believe they can be better explained in this analysis.

R2: The proposal should be likely to be effective;
R3: The proposal should be morally permissible;
R4: Among the proposals to be compared, select the one where the most grievous injustices are dealt with or, where this is not possible, seriously considered (Luna 2021: 25).

In his reconstruction, Simmons attributed the first three requirements to Rawls and he completed the Rawlsian proposal with the fourth requirement. Simmons' original requirement posited only the selection of proposals in which the most serious injustices were treated as a priority. In a later article (Luna 2021) I modify this final requirement and add a disjunction, since in many concrete situations, there is no possibility of giving priority of treatment, given the constraints of reality. In other words, this final requirement may be too strong. It may be that such a model, theory or proposal cannot directly respond to such injustices, but, for example, that they are considered for a later stage, or that conditions are created to deal with or avoid them. Another interesting point that Simmons mentions is that Rawls does not offer explicit criteria for weighing each requirement, nor does he clarify what the degree of satisfaction of each one should be. He points out certain inconsistencies (Simmons 2010: 19–20). In contrast to Simmons, I believe that this is precisely where its strength lies. While it is important to have a decision-making mechanism that ensures coherence in our reasoning and responses, I also maintain that the absence of strict 'order' or rigid criteria—as Simmons proposes—should be preserved, since each situation requires a dynamic and context-sensitive evaluation.

My 2015 article only considered the comparison between NITs. Now it is necessary to include ITs in the comparison. In the next chapter we will explore what IT implies (and theorisations that incorporate good idealisations or valuable idealisations *à la* Tessman). It should be borne in mind that we are working in bioethics, with relatively concrete situations, and where public policies may be implemented. And we are not dealing with the 'pure' ITs of political philosophy or ethics. While we need to arm ourselves with the proposals of political philosophy, our field of action is the real world in which there are no pure ITs but there are idealisations or public policies that may have a strong ideal content and that we can therefore consider as ITs. So, if we broaden our spectrum of possible theories (whether ITs or NITs), are these requirements a valid basis for comparison?

Note that the first two requirements deployed,

R1: The proposal should be politically possible;
R2: The proposal should be likely to be effective;

are directly related to feasibility, as there are different constraints and challenges in the real world. We need to consider the various constraints on feasibility[4]—from mild or contingent constraints such as cultural challenges, scientific findings or social values—to hard or unalterable constraints related to the laws of logic or physics. There may also be transitional constraints related to time. How then should we

[4] See Erman and Möller (2019) for a more comprehensive analysis of feasibility.

consider feasibility? Is infeasibility an obstacle to which we must surrender? How are we to deal with money, power and social position, among other constraints? These questions open up a muddy and complex field. Therefore, the first two requirements of our decision-making mechanism assess whether the actions to be taken are politically possible and whether they can be effective in this concrete world, in the face of the daily challenges that reality presents to us. In bioethics, it seems clear that we must incorporate these questions of feasibility or practicability and that we must include them in our normative deliberation.

The next two requirements of our decision-making mechanism are normative:

R3: The proposal should be morally permissible;
R4: Among the proposals to be compared, select the one where the most grievous injustices are dealt with or, where this is not possible, seriously considered.

The third requirement asks that the proposal must be morally permissible, i.e., it must take into account that not all the guidelines for action provided by NIT are acceptable, even if they are second best (Tessman 2010) or if they are viable alternatives. We should check what our assumptions are: whether our ideals or idealisations are appropriate[5] and we are not ignoring some relevant moral feature. Rawls is not too explicit about this and—as Simmons points out—leaves us to "guess at their meaning" (Simmons 2010: 20). I think it is important here to consider whether we are dealing with bad idealisations and, if so, to reject them. We should be able to justify which conceptual tools we are using, and we should carefully consider their real-world consequences. I propose a deliberative process: there are no rules or algorithms that can be applied to all situations and provide a clear answer. Through this process, for example, a rule, principle, strategy or policy that creates more oppression or increases systematic disadvantage for an already disadvantaged group should be avoided.

The fourth requirement must apply to the comparison between the different alternatives. So if we have a NIT1, a NIT2 and an IT, we should prefer the one that gives priority to resolving the most serious injustices or, if these cannot be resolved, seriously consider how to deal with them in the most appropriate way. This requirement allows us to critically re-examine bad idealisations that are blind to oppression or that assume ideal institutions without considering, among others, the systematic disadvantages of women, people with scarce resources or minorities. On the basis of this evaluation, the different theories can be rejected, prioritised or ranked. These last two requirements are closely connected.

In conclusion, the four criteria outlined in my 2015 article should be considered when selecting not only between NITs (Luna 2015) but also between NITs and ITs.

[5] We will develop this point in detail in the next chapter.

1.4 One Last Requirement?

But do we need other requirements to complement the above? If we think that IT may play a heuristic role that can help us not to accept the *status quo* or to have an inspirational impulse, we can incorporate another requirement based on the relevance of the final goal. This last requirement can reinforce consistency in our decision-making tool. The underlying idea is that while many NITs identify short-term reforms that seriously consider feasibility constraints that are linked here and now, longer-term proposals that may become relevant if the feasibility constraints are relaxed or changed should also be taken into account. And above all, that this evaluation should fundamentally assess that the proposed reforms or measures do not contradict or impede the achievement of the desired policy objectives. However, if we are to forego certain short-term benefits for the uncertain promise of greater long-term benefits, we must be cautious. Hamlin and Stemplowska advocate a genuine dialogue between IT and NITs, and do not simply posit IT as a prerequisite.[6] In other words, we should not assume a *priority claim* for IT or always consider IT as a trump card. The idea of dialogue is very interesting, as it is in line with the reflective equilibrium that I propose in the next chapter. We must be careful not to support actions or proposals that may hinder the achievement of long-term objectives.

Therefore, the following requirement holds:

R5: The proposal should appraise or consider long-term goals, in order to avoid endorsing strong obstacles to them when we try to move towards those goals (Luna 2021: 27).

The proposal can thus be illustrated in Fig. 1.1

As stated, a key role of this decision-making mechanism is to help us check for consistency and coherence in our advocacy for institutional and policy reforms, and to keep long-term goals in mind. Note, however, that in the selection and evaluation of ITs and NITs, we will have to filter out bad idealisations or assumptions. So the fifth requirement of the decision-making mechanism must consider how to make choices and trade-offs between different domains of partial justice and temporal trade-offs in terms of costs and benefits, and short- and long-term objectives.

1.5 Case 1: Conscientious Objection and Abortion

My article "Medical ethics and more: ideal theories, nonideal theories and conscientious objection" (Luna 2015) took the situation of legal abortion for rape and conscientious objection (CO) in Argentina before 2015 as a case study. At that time in the country, legal abortions were frequently denied and one of the causes was alleged

[6] Hamlin and Stemplowska, 2012, 59. I acknowledge that I used the priority assumption that Rawls proposed in my previous article (Luna 2021). However, I now believe that it should not function as an 'automatic clause' or have priority.

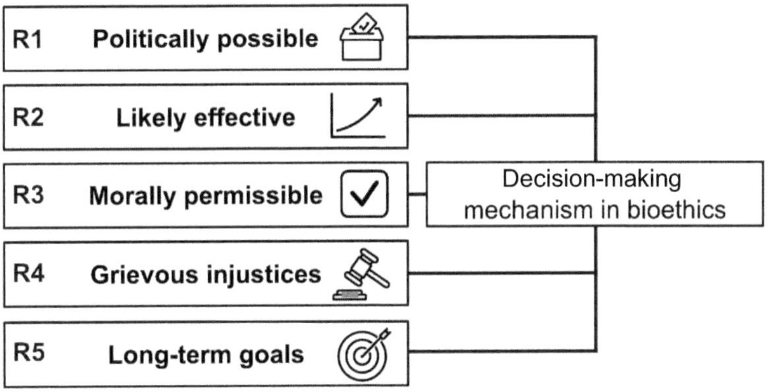

Fig. 1.1 Five requirements (R1 to R5) of a decision-making mechanism for bioethics to navigate degrees of ideality

CO. In December 2020, however, Argentina expanded the grounds for considering legal abortions. I assume—with a positive outlook—that the obstacles that existed before 2014 are now being modified and, as time goes by, will be resolved.

The article briefly explains how NITs can contribute to policy design in the non-ideal circumstances of conscientious objection (CO). I argued that in the reality of that time, NITs were preferable to realistically conceptualise problems arising in contexts of non-compliance and scarcity (Luna 2015: 129). In the ideal world (countries such as England or France were considered to be ideal worlds in the sense of this policy),[7] all interests and rights are respected and protected: those of the woman and those of the medical personnel involved. An outline of an IT on CO would involve the following prescriptions:

1. Permitted abortions must be performed;
2. CO must be accepted and respected.

IT for CO entails:

(a) that reproductive rights are respected,
(b) that services are accessible, and
(c) that CO is an exception.

These assumptions were not in place in Argentina in 2015.

Regarding points (a) and (b), Argentina, like many Latin American countries, did not offer a minimum standard of sexual and reproductive health care. There is a long history of denial of and disrespect for sexual and reproductive rights in the region that hinders safe and permitted abortions (Casas Becerra 2019). As a result, most abortions are performed unsafely and illegally, and involve harms and costs

[7] As has just been pointed out, in bioethics we work with the real world, which is obviously not ideal, but we can choose some situations (in this case countries) that are 'more ideal' in some respects, such as in respect for reproductive rights.

that women—especially those with fewer resources—have to suffer. Furthermore, there was almost no infrastructure in place to deal with legal abortions.[8]

If we consider point (b), we can see that in recent years it has become clear that permitted abortions must be performed by a public system which had no tradition in this respect. This situation is slowly changing.[9] In addition to individual reluctance, gynaecological services in the public health system presented "institutional objections", although it was mainly the Chief of Gynaecology who imposed his or her position and did not allow any professional of the team to perform abortions (Luna 2015: 130). CO and institutional objections[10] were used as weapons to ignore reproductive rights. In many cases, the same doctors who objected to performing abortions in public hospitals performed them in their private practices for a high fee. At the same time, the practice of CO had been completely abused: it did not reflect sincere beliefs and was used as a tool to refuse to abide by the few exceptions in the law that allowed abortions.

Given this situation, if CO was respected, an even greater lack of human resources could be expected, constituting a strategy to refuse to perform legal abortions. This also meant that fewer and fewer doctors would be available to provide the service. Moreover, those willing to perform them would be overburdened and could also be stigmatised. As a result, more poor women and women already at risk and facing the terrible situation of a pregnancy resulting from rape, had to resort to illegal, unsafe abortions, which meant a greater chance of harm or death, and re-victimised the victim. In this context, an attempt to impose the norm of perfect justice was likely to backfire, making the most disadvantaged even worse off. It did not follow, therefore, that we could simply apply IT (Luna 2015: 130). In Mills' words, this IT relies on idealisations to the exclusion or at least marginalisation of the actual (Mills 2005: 168).

1.5.1 Different Non-Ideal Possibilities

The three possible NIT options providing different solutions are:

(1) *Silent referral.* This position advocated an institutional policy of silence. The CO had to be reported to the head of the institution. Institutional objection (i.e., the entire gynaecology and obstetrics service refusing to perform abortions) was accepted. There was no challenge to the *status quo.* The only requirement was to have a plan to resolve the situation. Whoever ran the institution had to have a plan for referring the person in need of the service either to another person who did not object, or to another institution prepared to receive them.

[8] The practice of performing legal abortions hardly existed at the time of writing (especially in some provinces of Argentina).

[9] This has been the product of the long and difficult struggle of different feminist groups and now thanks to the new law decriminalising abortion. See Bergallo (2014).

[10] The accuracy of this terminology is a matter for discussion beyond the scope of this chapter.

The justification was that some reproductive rights had been achieved in that region, and if there was publicity, public records or any kind of debate, local religious power would prevail and turn back the hard-won gains. Everything would be handled without transparency or publicity.

(2) *Governance or careful regulation.* This position accepted neither silent referral nor institutional objection. It accepted referral as a compromise position and called for proper regulation of CO (i.e., developing a public and transparent register of objectors, establishing quotas of objectors in public institutions, etc.) and therefore set some limits to CO. It conceptualised objectors as privileged, as qualified health professionals and personnel and sought to change the *status quo.* This was the position of people living in big cities.

(3) *Temporary prohibition:* For this position, both institutional objection and CO should be prohibited in the public system. This should be maintained until there was true respect for sexual and reproductive rights and these services were actually provided by public institutions in a timely manner without re-victimising women. The backdrop to this position was the situation of total disrespect for reproductive rights in certain northern provinces that did not even comply with flagrant situations.

Note that the three options argue for different public policies that cannot be implemented at the same time (Luna 2015: 130–131). In that article I opted for the second NIT using the first four requirements presented, as it set boundaries, sought transparency and respected both CO and women.[11] It allowed for bypassing as a position of compromise. For example, careful governance or regulation determined quotas of objectors. This is key in public hospitals, as objectors may not be the right people for gynaecology wards or rural or small hospitals where a small number of teams provide most services. This implies that some of those who object may not be able to exercise their right if they wish to do so. Some restrictions therefore apply. They may be offered the option of practising another speciality, looking for a private clinic, or finding another job or hospital that still has vacancies for conscientious objectors. Careful governance or regulation required the existence of public registries to avoid the double standard of performing abortions as a private business, while denying them as a public service, and so that women knew which providers to access. It could also require objectors to make their views public (and thus avoid false objections).

1.5.2 How Ideal is Ideal?

Faced with this non-ideal scenario, let us consider whether IT can be pursued. To do so, I will apply the decision-making mechanism presented. First, we can argue that following the IT was not politically feasible (R1): conservative groups were so powerful that women's rights were not respected, as not every pregnant woman could access legal abortions. There are still multiple obstacles in many provinces in

[11] For more details, see Luna (2015).

Argentina today. But the *status quo* at that time was so strong that even undisputedly legal abortions were not performed.

Second, trying to follow IT will not respect requirement (R2) as it is ineffective: in the end, the abortion service will not be provided. It is precisely the aim of the objectors to put obstacles in the way of safe and legal abortion. Moreover, in some very conservative provinces no abortion service is provided at all. Such provinces only respect CO while continuing to ignore women. In extreme cases, anti-abortion groups occupied the hospital entrance and threatened women and their families. For example, in Tucumán in February 2019, an 11-year-old girl raped by her grandmother's partner was denied an abortion. The hospital delayed the procedure for almost two months because the entire medical staff of the public hospital were objectors. Despite an important civil movement calling for this girl's rights to be respected, it was impossible to change the dynamics and a Caesarean section was performed to protect the foetus, against the wishes of the girl and her family (López 2019).[12] This example—which occurred four years after my article (Luna 2015) was published—highlights how unrealistic it is to think that IT can be an effective guide to action. Furthermore, it illustrates the prevailing oppression suffered by women and pregnant women deprived of economic means who need public services.

In relation to the fourth requirement (R4), which states that "Among the proposals to be compared, select the one where the most grievous injustices are dealt with or, where this is not possible, seriously considered", we can argue that if we follow IT, this is not respected either. Women and pregnant women are on the weaker side of power inequalities and their rights are seen as secondary to those of the medical team. Moreover, in these cases, it is the poorest and most resource poor who generally need the help of the public system. Women and pregnant women with middle or high socio-economic status can access safe abortions even if they are illegal.

Those who support NIT will argue that doctors have a monopoly on certain practices and that their decisions to become health professionals are voluntary. They have been able to study, train and work. They come from a place of privilege and multiple possibilities, so society can and should establish rules and limits to their CO (Luna 2015: 130). Those who object can choose a specialisation other than gynaecology; they can work in the private sector or choose a hospital where there are vacancies for CO and the quota is not yet filled. In short, they have several options and, in this equation, they are in no way weaker than the women and pregnant women who need to terminate their pregnancies, were raped, and have nowhere else to turn (and let us remember, moreover, that these are legally accepted abortions). Therefore, these objectors may not always be able to exercise their CO as they wish, but it is morally permissible to pursue the proposed NIT. That is, NIT meets the third requirement (R3). Moreover, it can be argued that allowing IT will generate even more oppression and add ills or harms to an already disenfranchised group. The unjust and oppressive conditions faced by women and pregnant women must be factored into this equation. This is not a choice between equals. It is not possible to ignore this underlying situation by pretending that the autonomy and rights of women and pregnant

[12] Note the level of violence against women that such actions imply.

women are respected, nor to continue with unrealistic and idealised assumptions. This example clearly shows that there are questionable idealised approaches in IT and that IT cannot be pursued in the current imperfect conditions.

An NIT that sets some limits and safeguards would seem to be the second-best and most realistic prescription for action orientation. It is politically possible (R1) and effective (R2) in that it respects the rights both of women and of the medical team—even when it imposes limits on the rights of objectors. It is a morally permissible option (R3): it gives priority to the demands of women who need to terminate their pregnancies. As for the fourth requirement (R4), as we have seen, NIT addresses injustices by protecting women and pregnant women who need it most.

These four requirements were assessed in the 2015 article when deciding which NIT was the best. We should now consider what happens with our new requirement (R5): The proposal should consider or assess long-term objectives in order to avoid creating strong obstacles to them when trying to move towards them. In this respect, it could be argued that the three NITs presented seem to meet this requirement. While the temporary ban (3) appears to lay the groundwork for effective practice of the service, it does so by bypassing objectors and does not encourage greater coexistence between objectors and non-objectors. On the other hand, the silent review (1) respects everyone but does so in a shameful way and in secrecy. This does not help to create mutual respect and acceptance of the importance of undertaking these practices. In this sense, it is perhaps the most problematic because we do not know when these practices can be carried out without a problem, in broad daylight. Remember that these are legal services and there is no reason to hide them.

Considering R5, good governance or careful regulation (2) can be advocated since, even if it imposes some limits on practitioners, the long-term goal is to respect the rights of both parties. And if we enforce this policy, it defines a path towards a more respectful attitude on all sides. It also helps to build an infrastructure providing legal abortions, as they will begin to be performed systematically. Those who object and are in institutions or services where they can be replaced in the performance of the practice will be replaced. And all public institutions will be able to provide the service because quotas for objectors will allow it. The proposed NIT policy meets the fifth requirement, avoiding future obstacles to IT. Instead, it fosters awareness, infrastructure and respect for pregnant individuals—including women and other gender identities—as well as for objectors. Note also that if, indeed, the situation does change and the country improves in its infrastructure as well as increasing respect for these women and their rights, the transition to an IT may well be possible. The selection of the NIT may be temporary as the situation may change and allow for another type of policy that effectively respects all the rights involved and is in line with the IT. In this sense, of the three NITs, the path of governance and regulation is perhaps the most appropriate.

1.6 Case 2: Research with Pregnant Women

Our second case is also related to reproductive rights. After the thalidomide case in the 1960s,[13] very little pharmacological research was conducted with pregnant women. In recent years, however, the need for research with this population has been recognised. The responsible inclusion of pregnant women in research was the focus of the Second Wave Initiative launched at Georgetown in 2009 (Heyrana et al. 2018). Despite this, there remains much resistance, from funders and researchers alike, to including pregnant people due to the possibility of harming the foetus or triggering costly lawsuits. However, sick people fall pregnant and pregnant people fall sick, and more than 80% of pregnant women and pregnant people are prescribed medicines that have not been studied during pregnancy (Ayad and Constantine 2015). According to the US Department of Health and Human Services, of the 213 new drugs that received FDA approval from 2003 to 2012, only 5% included data from pregnant women. These data show how supposed protections end up leaving this population unprotected in the long run.

1.6.1 An Ideal Proposal

The need to conduct such research has been underlined and supported by ethical guidelines, such as the CIOMS *International Ethical Guidelines for Biomedical Research Involving Human Subjects.* The *Guidelines,* which are endorsed by the World Health Organization (WHO), are global in scope and therefore have great international impact. Guideline 19, which refers to pregnant women and pregnant persons, clearly explains which ethical elements should be considered when conducting research on pregnant women, highlighting the need to promote this type of research and that it should only be initiated after careful consideration of the best available data. It also makes explicit safeguards, such as that the informed consent must be that of the woman participating in the research.[14] It considers the potential benefit to the pregnant person or her foetus, and also how to deal with research interventions without potential benefit. Interestingly, the Guideline ends by noting:

As a general rule, health-related research involving pregnant women that has the potential for harm to the fetus should be conducted only in settings where women can be guaranteed access to a safe, timely and legal abortion in the event that participation in the research makes the pregnancy unwanted (CIOMS 2002).

---

[13] This drug was administered to pregnant women without prior research. Many children were born with serious problems and malformed limbs. Another example is DES or diethylstilbestrol, a synthetic estrogen used between the 1940s and 1970s to prevent miscarriages (https://www.cancer.gov/espanol/cancer/causas-prevencion/riesgo/hormonas/hoja-informativa-des).

[14] This recommendation, which is obvious in some countries and contexts, is not obvious in others where the family or the husband takes priority in the decision.

Guideline 19, while encouraging this type of research, sets out the requirements for good research conditions that respect the value of science, researchers and pregnant women. We can consider this Guideline as a kind of "ideal theory" or an idealised proposal, for two fundamental reasons. First, because it protects pregnant women and prevents them from being harmed (and, at least in Latin America, there is a tradition of abuse and mistreatment of women in the area of sexual and reproductive rights). In addition, malformations of the foetus or other serious problems may occur and the pregnant woman may not want to continue with the pregnancy, which may cause health complications or even fatalities in contexts where abortion is not legal, safe and accessible. Note that the requirement is quite strong as it calls for abortion to be not only safe, but also "timely", i.e., performed legally in a timely manner (which implies a certain haste). Second, this proposal promotes respect for reproductive decisions. Although it considers the non-ideal world in which abortion is not performed legally and safely, it recommends a rule to be followed in line with certain principles (i.e., it protects the woman and the pregnant person and avoids exposing them to unsafe abortions).

The phrase "in the event that participation in the research makes the pregnancy unwanted" establishes an ideal-type restriction in the non-ideal world. In countries where abortion is prohibited, termination even of a desired pregnancy may be required, due to possible teratogenic effects or malformations in the foetus caused by research.[15] This restriction applied in Argentina until December 2020, although now the situation is less clear because the law only permits abortion up to a certain number of weeks.[16] Most Latin American countries where this type of abortion is not legal are also affected, and following the reversal of the Roe vs Wade ruling in the USA in 2022 several states went from full reproductive rights to extremely restrictive policies.[17]

1.6.2 A Non-Ideal Proposal: Optimistic Science

Although there are several types of NITs, the one proposed here is what I call "optimistic science". The proponents of optimistic science emphasise the value of research and view barriers as challenges to be addressed so that studies can be conducted and scientific progress is not halted (Luna 2007a; 2007b).

Research ethics are permeated either by more idealistic or by more realistic and pragmatic positions. The use of placebo as a comparator illustrates these two positions. Generally, research is designed to compare outcomes, so a new treatment or

---

[15] That is, when abortion for foetal malformations, for problems affecting the health of the pregnant person and on demand are illegal or not accessible.

[16] Law 27610 allows the termination of pregnancy up to and including the 14th week without the requirement to justify it, from the 15th week onwards it must be due to rape or if the life or health of the pregnant woman is in danger. In other words, in Argentina a woman is only fully covered if the abortion can be performed before the 15th week.

[17] See the article by Sugarman et al. (2023) on the problem of how to do research.

drug is tested against the best existing treatment, and when there is no treatment, it is compared against a placebo (usually a harmless substance). Advocates of the NIT of optimistic science generally argue for the use of placebo if there is a treatment that is too expensive or difficult to implement, since in general, using placebo means fewer research subjects, less expense and purer data. However, if there is a treatment, people who need it will not receive it. The discussion is in fact very complex,[18] as there are various circumstances that can ethically justify the use of placebo (there are those who raise methodological reasons, those who accept some types of psychosomatic illnesses that do not involve serious risks, as subjecting a person to harm when there are effective treatments cannot be justified). In the late 2000s, widespread debate on this issue led to changes in major ethical documents such as the Helsinki Declaration and the CIOMS guidelines. The simplification presented here merely illustrates a theoretical position based on more ideal principles versus one that is not, in order to demonstrate the approach of those who defend the non-ideal position of optimistic science.

Thus, IT proponents will not accept placebo as a comparator except when there really is no treatment (for example, at the beginning of HIV-AIDS in the 1980s when it was not known how to fight the disease and there were no treatments) or in a few other exceptional cases.[19] This proposal is based on "principles" and will call for research to be stopped when it does not respect certain basic issues. In fact, when AZT and placebo research to prevent perinatal transmission of HIV/AIDS in Africa was denounced, guidelines such as the Helsinki Declaration prohibiting the use of placebo were revised. Angell (1997) and Lurie and Wolf (1997) condemned the use of placebos in such research, since at that time a proven effective treatment already existed. This position was considered too idealistic and not pragmatic, as it set up further barriers to an already difficult and complex activity.

A reconstruction of these debates is beyond the scope of this chapter,[20] but it shows how two positions in research ethics emerged, one more idealistic and the other more pragmatic, advocating different scientific standards for conducting research. Following the latter, in the present case we can imagine NIT proponents questioning the policy set out in CIOMS Guideline 19 as being too protective and arguing instead that research can be conducted despite some risks (in the same way as they argued for placebo).

So, the main arguments that can be put forward by those who defend the NIT of optimistic science in the case of research with pregnant people could be:

(a) clinical trials are carefully designed to minimise risks;
(b) failure to conduct such research implies that pregnant women or pregnant persons who would have participated in the research will not derive a potential benefit;

---

[18] This is a controversy in research ethics that could already be considered 'classic'.

[19] Although this is a discussion from the late 20th and early twenty-first century, it is still relevant. This was the convening theme of the Helsinki Declaration discussion in Sao Paulo in February 2023.

[20] For more on this issue see Luna, 2001 and Luna, 2007a and 2007b.

(c) this position not only has a negative individual impact, but also has negative collective effects as it slows down scientific and technological progress; and
(d) the rights of women or pregnant women are respected through informed consent.

Thus, advocates of NIT can defend their position by pointing out that the participants freely give their consent and that they know clearly what they are exposing themselves to and accept it. If they wish, they can have a disabled child rather than risk an illegal and unsafe abortion. They can also argue that, ultimately, these women and pregnant persons may have an unsafe and illegal abortion (as most women in these countries do). This is the *status quo*. The only restriction NIT will have to accept is that only research with the potential to benefit (women, pregnant women or foetuses) can be done. Even for those who advocate a positive view of science, rejecting this restriction—although it could advance knowledge—seems unacceptable and even cynical. It does not meet the minimum requirement of care that exists for any research. Therefore, I do not consider it an option for discussion. For the NIT of optimistic science, only potentially beneficial trials should be accepted—even if they may expose women or pregnant women and their foetuses to teratogenic effects—in countries where abortion is illegal and they have no access to a safe termination of pregnancy.

1.6.3 A Hypothetical Situation

Pregnancy and HIV share a risk factor: unprotected intercourse puts persons and pregnant women at risk of infection. Negotiating condom use is more difficult for some people.[21] In addition, when pregnant women acquire acute HIV infection during pregnancy, rates of vertical transmission, i.e., transmission from mother to foetus, are significantly higher than before pregnancy.[22] Oral pre-exposure prophylaxis (PrEP) has been shown to prevent HIV infection in high-risk populations (e.g., men who have sex with men, people with addictions, sex workers), but little is known about its use in pregnant women. Taking a daily pill has been shown to be highly preventive.

Data and drugs (and ways of taking them) are changing. Because of these changes it is preferable to take this case as hypothetical, as at the time of publication, the drug of choice is perhaps not the right one. But the problem is clear: with which population should a first clinical investigation be carried out in pregnant people that may cause malformations or problems in the foetus? The fundamental point here is to assume that there is no clear data yet on the health impact on the pregnant person and her foetus.

Proponents of NIT may argue for the inclusion of particularly at-risk pregnant women in these trials. Therefore, as an exercise in argumentation, let us assume

[21] Especially in some low- and middle-income countries, where men are reluctant to use condoms.

[22] In a study of nearly 10,000 women in South Africa, among those who seroconverted during pregnancy, 11% transmitted the disease to their children, compared to 2% who were HIV positive before pregnancy (PHASES 2017).

that the drug to be tested has the potential to produce teratogenic effects and that there are as yet no reliable data. There is animal testing and some encouraging data in pregnant people who unknowingly took these drugs without experiencing major adverse effects. This is important research and may provide relevant knowledge. But the drug has not been tested before in pregnant people, there are no clinical trials and/or there is no solid evidence of its effects on the foetus. In fact, this is often the case because for research to be justifiable there must be uncertainty about the effectiveness and adverse effects of the drug or treatment in question. This is precisely why research is conducted. If it is known with certainty that the new treatment or drug is better than what exists, the research should be terminated immediately and treatment standards modified. Therefore, there is always a halo of uncertainty to justify initiating research. Note that in this particular case, this is the first controlled research in pregnant people. Once there are data revealing that there are no teratogenic effects or serious adverse effects, the research should no longer be problematic. It is precisely this first decision making that we are working on in this hypothetical case.

Presented with this case, supporters of IT will refuse to carry out such research in Argentina[23] and in several Latin American countries even if it has a valuable perspective. They will argue that, despite their importance, these trials should first be conducted in countries where abortion is safe and accessible (that is, where "eugenic abortion" is legal and accessible, such as France, the UK and Switzerland). And, once data on teratogenic effects are obtained, research in these other populations may be permitted. It should be noted that this case does not concern individuals or couples who do not want to continue the pregnancy but who, due to the possibility of malformations or serious effects on the foetus, consider this possibility as an alternative. The IT position is not against all research. Advocates of IT see this clause as a requirement, as a general rule to protect these people, so that only very specific and well justified exceptions are acceptable. In the case of Latin America, Zika research may be an exception that IT would accept. Countries in the region, most notably Brazil in 1916 (PAHO-WHO 2016). In such a case, research with new drugs that is the first research on pregnant people cannot be done in other parts of the world that are not affected by the disease in question, and it is foetuses in Latin American countries that are particularly affected. In Africa the same reasoning could be applied to Ebola: in the Ebola epidemic of 2014 no foetuses survived with respect to maternal Ebola and a higher proportion of pregnant women infected with Ebola died (Heyrana et al. 2018). In these very particular situations, not doing research will leave these populations completely unprotected, as research cannot be done in England or France where the disease does not exist. Therefore, IT, while accepting some very well justified trials, will undoubtedly be more restrictive than NIT, hindering the hypothetical research we suggested.

[23] Assuming that the effects can be known from the 15th week onwards (see quote 16).

1.6.4 The Decision-Making Mechanism in Practice

We have seen how the protective position of IT rejects the need to engage in such trials, while NIT's optimistic view of science will defend it. How do we decide which is the appropriate response? Let us consider our decision-making mechanism to assess whether IT or NIT should be followed.

The first requirement (R1) asks whether this NIT is politically possible. From the optimistic science point of view, it appears to be the case. The countries in question operate without regard for the sexual and reproductive rights of women and pregnant women, as has just been argued in the previous case, and for decades even legally permitted abortions were not performed, so this is part of standard practice. This is in line with many other policies that these countries accept. For example, there are legal and accepted prenatal tests that help couples know how their baby is progressing.[24] In the event that these tests show malformations or serious problems, some couples—a minority—will accept this fact and be better prepared for a difficult birth, but the majority of couples—referred by their doctors and physicians—will secretly and illegally terminate their pregnancies. This is common knowledge and does not imply any political problem. It is part of standard practice in countries where there is widespread disregard for the sexual and reproductive rights of pregnant women.[25] In the case of these pregnancy trials, advocates of NIT will argue: we can benefit some of the pregnant women, and only those who may suffer teratogenic effects or who are so distressed that they do not want to continue the pregnancy will be treated unfairly or harmed. These people will do what most pregnant women in these countries do: depending on their financial means, they will seek an illegal and, in most cases, unsafe abortion. In our hypothetical case, the scientific reasoning will be to start with the population that is most at risk, so the research will begin with a population composed of high-risk individuals. If we return to our hypothetical case of prophylaxis for HIV-AIDS infection, we will be dealing with drug addicts, sex workers, etc. They are the ones who are most in need of this preventive treatment; they are also likely to lack financial resources and will therefore end up having not only illegal but also unsafe abortions. However, advocates of NIT will argue that this is not a problem: illegal abortions are the only viable solution in these contexts. This is the *status quo*, the baseline for these populations. As will be argued below, this baseline is too questionable to be taken as unproblematic.

In relation to the feasibility constraint, advocates of NIT will argue that the second requirement (R2) asks whether NIT is likely to be effective and their answer will be that it can be. Conducting these trials will provide the knowledge they seek. They will advance the science and, in principle, will benefit all pregnant people (whether

---

[24] This is the case in Argentina and many other countries.

[25] In Argentina this has been happening for decades without any problem. A similar situation arises with assisted reproduction techniques and multiple pregnancies. Abortions are performed without any problem when women have multiple pregnancies. In these cases, the already implanted embryos are removed to leave only one or two embryos in order to ensure that the pregnancy is successful, as a pregnancy with quintuplets or even triplets is highly problematic (Luna 2018).

the results are positive or negative), and especially the women who have participated in the trial. It is true that no matter what the outcome is (positive or negative), it will provide knowledge and therefore have scientific value and contribute to science. However, advocates of NIT will have to admit that they will benefit these particular women only if PrEP proves to be safe and effective in the trial; but if the drug has teratogenic effects, it may harm these women and their foetuses. Moreover, it may lead to the death of some women through unsafe abortions (especially those that cannot be performed pharmacologically in relatively advanced pregnancies). This situation may be very likely due to the timing of the research. So R2 can only be justified for participants if the results are positive, but this is precisely what cannot be assured prior to the trial in question.

The third requirement (R3) of the decision-making mechanism asks whether this NIT is morally permissible. Those who argue for NIT will argue that informed consent is of paramount importance. It must be done with care; it must provide all information, including the possibilities of harm to the foetus during the research. If uncoerced pregnant women consent, they should be respected. It is their autonomous choice. Proponents of NIT may argue that this research offers these people the possibility of HIV prevention (in other cases it may offer cures for diseases, help for their foetuses, benefits they would not have enjoyed if they had not participated in the research). They will argue for and emphasise the possibility of the benefit. It is not that they are intentionally discriminating against such people, but on the contrary, they are trying to help these already disadvantaged pregnant women who—in addition to having a difficult life—are exposed to HIV/AIDS. They may also argue that this is the starting situation, the baseline, as is the case in other countries where health care services function poorly and where research is currently allowed.[26] In these countries the baseline situation involves unsafe and illegal abortions, so even with risks, it is preferable to move forward with these trials rather than conduct no research at all.

Conversely, those who defend IT will argue that this NIT cannot be accepted as morally permissible. In this case the baseline risk appears to be too high and above the normative proposal: it is not a minimal risk or a minor increase above the minimal risk. An unsafe and illegal abortion can harm or kill the woman, or in some extreme situations send her to prison.

IT proponents may also argue that informed consent has never been the only safeguard that has been considered in research ethics and cannot be taken as the sole or even the most important criterion. Moreover, a fundamental element in the assessment of a research protocol is the risk to which the subjects are exposed. Research ethics reminds us that informed consent is a necessary but not sufficient condition (Luna 2008). Moreover, one can go further and question the validity of informed consent in contexts where there is no real choice: how voluntary is the participation in such research of a sex worker or an addict when, if the trial has problems, this will imply the possibility of an illegal and unsafe abortion?

[26] Here the similarity with the arguments in favour of placebo use is striking.

As we have already seen, feminist positions can also argue for a more sophisticated view of autonomy that includes the impact of context and relationships, such as relational autonomy (Buedo and Luna 2021). Those who argue for NIT overlook women's oppression, lack of resources and the pressure they may feel to participate in the trial. These are people at high risk of HIV infection, such as addicts or sex workers, who are also discriminated against and stigmatised. Note that, in this case, it is the NIT advocate who is using bad idealisations: persons (in this case women) as perfectly rational decision-makers (which shows that bad idealisations do not only belong to the realm of the ideal). Furthermore, those who advocate for IT can argue the need for change and the importance of respecting sexual and reproductive rights. They can give reasons for the goals of equality and respect for sexual and reproductive rights to which they aspire. In this sense, IT can indeed advocate for changes—in this case changes in sexual and reproductive health policies and laws—so that safe research can be conducted in the region. As we will see, it would imply what Gilabert calls dynamic duties (Gilabert 2017).

In this discussion it does not matter how careful and serious the researcher is—as noted above, a clinical trial *itself* raises uncertainties. Therefore, those who advocate IT will take the side of the research participants, while those who follow NIT will prioritise the value of the science, the research and its potential benefits.

Can we use other criteria to help us decide which proposal to choose? The fourth requirement (R4) of the decision mechanism states that the most serious injustices should be given priority treatment. Are we respecting R4 with our NIT? Let us revisit the justification framework for the four NIT arguments formulated in the section "A non-ideal proposal: optimistic science" and assess how it works.

Argument (a) points out that one must have a careful design to try to minimise risks, but there is no certainty that all risk (or at least teratogenic effects) can truly be avoided, just as there is no certainty of benefit. These are basic characteristics of any clinical trial. As far as argument (b) is concerned, the IT advocate will have to accept that these particular women will not have the opportunity to participate in the trial, so will not gain any potential benefits. This seems to be the strongest point of the NIT advocate. In the face of this retort, the IT advocate may argue that there may be other ways to obtain the drugs, such as through an innovative practice or a non-validated practice if the first positive data are obtained (Holzer and Mastroleo 2019).[27] However, this does not seem a conclusive counter-argument. Argument (c) is based on the progress of science. Science can advance with the knowledge gained from other, better protected participants (in countries where abortions can be safely performed). IT advocates are not against conducting such studies; they are only against conducting them in populations who cannot be adequately cared for. Thus, if we have to choose between two groups to test a drug, following the principle of risk minimisation, we should select the one that exposes the population to less risk if the information obtained is the same. Finally, we have seen that argument (d) based on informed consent is not so strong. We have already analysed the incorrect idealisations it entails. Moreover, although this last argument may allude to respect

[27] I thank Ignacio Mastroleo for this suggestion.

for women's wishes and choices, this does not seem to be the case in these stigmatised and disregarded populations.

On the other hand, it seems that the most serious injustice is the lack of substantial respect for the reproductive rights of women and pregnant women. This may involve subjecting them to unsafe and illegal abortion that may result in harm, morbidity or mortality. This is unacceptable, especially if such research can be conducted in places where reproductive rights are respected and pregnant women and pregnant persons can participate safely. In this sense, the NIT arguments based on the value of science and possible individual benefits do not overturn the lack of respect for reproductive rights.

Finally, let us examine the fifth requirement (R5): Does NIT's proposal—while upholding the value of research and science—assess or weight long-term goals? The NIT appears limited to short-term considerations: only the set of pregnant people who would have participated. They would be denied the possibility of hypothetical benefit (the real benefit is not certain). This corresponds to argument b), which we have already pointed out is the strongest argument in NIT.

However, one counter-argument that IT advocates can present is the following: the implications of such policies must be assessed, particularly the risks associated with the research. That is, how should the situation be viewed if there are indeed teratogenic effects and babies are born with serious problems or pregnant people die from unsafe abortions as a result of having participated in research? This can be analysed by considering two issues: (i) the research activity itself, and (ii) reproductive rights.

As for the first, it appears that we can conduct trials in these countries under conditions that are unacceptable in other countries. Advocates of IT may denounce a double standard in research. Remember that there are many countries that do accept and provide these abortions legally and safely. Such trials may even show a certain selfish interest or reflect the "greed" of people who conduct this type of research to benefit themselves. Moreover, as some Latin American countries have resisted certain clinical trials, this policy may be counterproductive to the good research effort by pointing to unethical behaviour that is unnecessarily conducted in countries that do not have adequate protections. This can lead to a stricter stance against research. It can promote the creation of overly protective laws that can hinder good research.[28] Strategically, therefore, it is a rather questionable policy that can have serious and negative repercussions for research. This is without even considering whether the alleged research would have caused morbidity or mortality in this population and would be picked up by the press. In the latter situation the damage that such research, based on the NIT model, would cause would be immense and would affect all local

[28] Note that the highly restrictive position of distrust of research has been prevalent in the region. Costa Rica banned research for five years. See "Costa Rica approves regulations to restart human medical research". Retrieved June 12, 2019, from The Tico Times Costa Rica website: https://ticotimes.net/2015/07/22/costa-rica-approves-regulation-restart-biomedical-research-humans. See also Mora Espinoza (2017).

science and research. Based on the above, not only the short-term but also the long-term consequences should be considered. As well as the model of science to be defended: a science that respects the people who collaborate with it.

With regard to reproductive rights and gender, the situation shows a generalised lack of respect for women and pregnant women. IT advocates may argue that allowing such research reinforces the denigratory and disrespectful *status quo* with respect to reproductive rights. Note that a recurring issue is whether or not to accept the *status quo*. From a theoretical point of view, both IT and NIT can challenge the *status quo*. A position such as that of Mills (2005) argues against the *status quo* (highlighting assumptions or silences of idealised positions in the face of oppression)[29]; but from the point of view of idealised or utopian positions, one can also confront the *status quo* by seeking to respect ideals or advocating for a better society. Thus, the IT advocate will argue that acceptance of these trials will prolong a history of reproductive rights violations. Without acknowledging the harms of supporting the NIT policy, people will continue to deny and ignore the plight of millions of women and pregnant women.

Opposing these trials, on the other hand, can help reveal the widespread lack of respect for reproductive rights; this needs to be changed so that good clinical trials can be conducted in the region (note that this is quite idealistic and is a long-term goal that could not be changed in the short term). In fact, researchers can argue for the need to change the law. For example, in Argentina, in 2018, a bill to change the abortion law was introduced in the Chamber of Deputies, which was subsequently rejected by the Chamber of Senators. Professionals from different fields, including health professionals, advocated for and against the bill in hearings held by both Houses. So researchers can also advocate for appropriate abortion laws. By bringing together researchers, doctors from different specialties (obstetrics, genetics, gynaecology) and other professionals, the importance of a law that respects sexual and reproductive rights can be demonstrated. Respectful reproductive laws will allow Latin American countries to participate in the same type of research as other developed countries (with high levels of protection).[30]

The latter arguments take into account long-term goals and reject the *status quo*. So in this case, it may be better to choose IT, which restricts this type of research, at least temporarily, and seeks to generate real protection of reproductive health and an ethical and acceptable way of doing research. The IT position does not imply an obligation on the part of those doing the research to argue for or have to carry out reforms themselves, although by refusing to carry out the research they expose local shortcomings or problems.

This case shows how a more ideal position is acceptable, and also that not all NITs are robust and valid; some may be very partial and focused on a few feasibility constraints or, even if non-ideal, may involve questionable idealisations (such as informed consent being truly voluntary). As Tessman (2010: 797) points out, there

---

[29] See the next chapter of this book, especially the section on the ideal versus non-ideal.

[30] Remember that this is the situation in most industrialised countries: the United Kingdom, France, Switzerland, etc.

may be NITs that only guide action—very strongly adapted to the non-ideal—and although they may be the best possible, they would nevertheless be unacceptable. We can formulate several NITs and not all of them will have the same value. ITs may also use bad idealisations or ignore background conditions. Therefore, there is no automatic answer as to which of these theories is the best. We need to look case by case, to examine whether bad idealisations are being used, whether the most serious injustices are being considered, and how they are interacting with the context, whether worthy ideals are being included, and so on. But such an assessment cannot be arbitrary. In this sense, having the five proposed requirements allows us to take a first step towards showing that these are not merely contradictory responses, and it helps us to avoid arbitrariness.

1.7 Conclusion

In this chapter, I have developed a decision-making mechanism to evaluate ITs or NITs, expanding it from its initial four requirements to five. I analysed two cases in detail and put the decision-making mechanism into practice.

In the case of conscientious objection, IT assumes a supposedly functioning society—a society fully committed to reproductive rights—rather than recognising the strength of resistance and the systematic disrespect towards women and pregnant people that in reality persists. IT is blind to what is really happening and to ways of rectifying these unjust situations. Following an IT in this context would have harmed those who were weakest and would not have promoted long-term goals.

In contrast, in the case of research involving pregnant people, IT protects the most vulnerable, proposes long-term goals and seeks to change the *status quo*. Therefore, before accepting NIT in this case, we must bear in mind that what is achievable in the real world may not be good enough, and that there may also be worthy ideals or long-term goals to consider. Why do we need this complex process of analysis? Because sometimes we need to use an IT, while at other times the best solution relies on a NIT. Responses are not automatic. Deep deliberation and a degree of consistency in reasoning are required. Thus, some of the questions we must ask are: How can these different theories and answers be understood? How can an IT and a NIT be conceived? Are there positive elements in each theory to consider? What is their theoretical basis? How should bioethics proceed? What assumptions are at work? The next chapter will deal with the theoretical underpinnings of this analysis of concrete cases.

References

Angell, M. 1997. The ethics of clinical research in the third world. *New England Journal of Medicine* 337 (12): 847–849.
Ariza, S., and L. Saldivia. 2015. Matrimonio igualitario e identidad de género sí, aborto no. *Derecho y Crítica Social* 1 (1): 181–209.
Ayad, M., and M. Constantine. 2015. Epidemiology of medications use in pregnancy. *Seminars in Perinatology* 39 (7): 508–511.
Bergallo, P. 2014. The struggle against informal rules and abortion in Argentina. In *Abortion law in transnational perspective: Cases and controversies*, ed. J. Rebecca.
Buedo, P., and F. Luna. 2021. Toma de decisiones compartidas en salud mental: Una propuesta novedosa. *Revista De Medicina y Ética* 32 (4): 1087–1110.
Casas Becerra, L. 2019a. ¿Todos ganan con la ley de aborto en Chile? De la objeción del stock holder a la objeción institucional SELA. Working document.
Casas Becerra, L. 2019b. La academia entre 'los seculares y los confesionales'. Las continuidades en la reforma por la despenalización del aborto en Chile y una historia personal. In *Aborto en tres causales en chile. Lecturas del proceso de despenalización*, ed. L. Casas Becerra and G. Maira Vargas, 203–233, Centro de Derechos Humanos, Facultad de Derecho, Chile.
Cook, R.J., Joanna N. Erdman and Bernard M. Dickens. 2014. *Abortion law in transnational perspective: Cases and controversies*, 143–165. Philadelphia: University of Pennsylvania Press. https://doi.org/10.9783/9780812209990.143.
Council for International Organizations of Medical Sciences (CIOMS). 2002. (2016). *International ethical guidelines for health-related research involving human subjects*, 3rd ed. Geneva: CIOMS.
Erman E. and N. Möller. 2019. A world of possibilities: The place of feasibility in political theory. *Res Publica* 1–23. https://doi.org/10.1007/s11158-018-09415-y.
Gilabert, P. 2017. Justice and feasibility. A dynamic approach. In *Political utopias. Contemporary debates*, ed. M. Weber and K. Vallier, 95–126. Oxford University Press.
Hamlin, A. and Z. Stemplowska. 2012. Theory, ideal theory and the theory of ideals. *Political Studies Review* 10 (1): 48–62. https://doi.org/10.1111/j.1478-9302.2011.00244.x. *Journal of Medicine* 337 (12): 847–849.
Heyrana, K., et al. 2018. Increasing the participation of pregnant women in clinical trials. *Journal of the American Medical Association* 320 (27): 2077–2078.
Holzer, F. and I. Mastroleo. 2019. Innovative care in Latin America: definition, justification and ethical principles. In *Controversies in Latin American bioethics*, ed. M. Hevia and E. Rivera López, 145–176. Springer International Publishing.
López, F. 2019. Denunciaron penalmente a los médicos que le hicieron una cesárea a la chica violada en Tucumán. *La Nación*, 11 March. https://www.lanacion.com.ar/sociedad/denunciaron-penalmente-medicos-le-hicieron-cesarea-nena-nid2227505.
Luna, F. 2001. Is 'best proven' a useless criterion? *Bioethics* 15 (4): 273–288.
Luna, F. 2007a. Experimentación con Seres Humanos. In *Nuevos Materiales de Bioética y derecho*, ed. M. Casado, 177–194. Fontamara.
Luna, F. 2007b. Research in developing countries. In *The oxford handbook of bioethics*, ed. B. Steinbock, 621–647. Oxford University Press.
Luna, F. 2008. Informed consent: still a useful tool in research ethics. *RECIIS. Electronic Journal of Communication Information & Innovation in Health* 2 (1): 41–51. http://www.reciis.cict.fiocruz.br/index.php/reciis/issue/view/15.
Luna, F. 2015. Medical ethics and more: Ideal theories, non-ideal theories and conscientious objection. *Journal of Medical Ethics* 41: 129–133.
Luna, F. 2018. From the Middle Ages to the 21st Century. Abortion, assisted reproduction technologies and LGBT rights in Argentina. *Canadian Journal of Bioethics* 1 (2): 26–36.
Luna, F. 2021. Ideal and non-ideal theories: The challenges of decision-making in an imperfect world. In *Applying nonideal theory to bioethics: Living and dying in a nonideal world*, ed. E. Victor and L. Guidry-Grimes (139), 17–40. Philosophy & Medicine Series, Springer.

Lurie, P. and S. Wolfe. 1997. Unethical trial interventions to reduce perinatal perinatal transmission of the human immunodeficiency virus in developing countries. *New England Journal of Medicine* 337 (12) (September 18): 853–856. https://doi.org/10.1056/NEJM199709183371212.
Mills, C. 2005. 'Ideal theory' as ideology. *Hypatia* 20 (3): 165–184.
Mora Espinoza, M. 2017. La investigación clínica en Costa Rica se encuentra en un punto crítico: Aplicación actual de la 'Ley Reguladora de Investigación Biomédica.' *Acta Médica Costarricense* 59 (3): 88–89.
Panamerican Health Organization (PAHO)—World Health Organization (WHO). 2016. *Zika ethics consultation: Ethics guidance on key issues raised by the outbreak.*
Pregnancy +HIV/Aids Seeking Equitable Study (PHASES). 2017. Ethical inclusion of pregnant women in the HIV research agenda. Working document.
Rawls, J. 1999. *The law of peoples*. Cambridge: Harvard University Press.
Robeyns, I. 2008. Ideal theory in theory and practice. *Social Theory and Practice* 34 (3): 341–362.
Simmons, J. A. 2010. Ideal and nonideal theory. *Philosophy & Public Affairs* 38 (1): 5–36.
Sugarman, J., et al. 2023. Ethical research when abortion access is legally restricted. *Science* 380 (6651): 1224–1226.
Tessman, L. 2010. Idealizing morality. *Hypatia* 25 (4): 797–824.
US Department of Health and Human Services. 2009. Transmission of the human immunodeficiency virus in developing countries. Protection of human subjects. https://www.hhs.gov/ohrp/sites/default/files/ohrp/humansubjects/regbook2013.pdf.pdf.
World Medical Association (WMA). 2013. *Declaration of Helsinki. medical research involving human participants.*, 64th General Assembly, Fortaleza, Brazil, October 2013.

Chapter 2
Theoretical Analysis—Meta-Bioethics?

Abstract The second chapter goes into detail with ideal and non-ideal theories. It deals with analyses of political philosophy and translates them into bioethics. It reviews the different conceptual distinctions and some of the debates and criticisms of the ideal and the non-ideal. It justifies the decision-making mechanism developed in the first chapter because when it is assumed there is a multiplicity of theories with a different degree of ideality that are valid and that there are valuable elements we can rescue from the ideal and non- ideal proposals; a decision- making mechanism such as the one proposed is needed. This chapter makes explicit some of the theoretical elements presented or suggested in the first chapter. This is the book's most theoretical chapter, with a meta-bioethical focus, adapting political philosophy's debates to bioethics.

Keywords Ideal and non-ideal theories · Decision-making mechanism · Dynamic obligations · Non-ideal concepts · Meta-bioethics

2.1 Introduction

The non-ideal world in which we live is marked by oppression, corruption and injustice, with individuals frequently failing to meet their obligations or to respect basic rights. In Chap. 1 we looked at the assumptions and theoretical elements that underpin bioethical reasoning. The analysis demonstrated that, in one morally significant case, it was necessary to rely on an ideal theory (IT), while in another, a non-ideal theory (NIT) proved more appropriate. But how can this be the case?

This chapter aims to clarify the relationship between ITs, NITs and bioethics, while also analysing some crucial concepts, such as vulnerability and autonomy, and showing their connection to the non-ideal. To achieve this, I take a step back to reflect on the theory underpinning practice and the assumptions that are implicitly involved. The chapter explores the challenges we encounter when attempting to apply

Based on the article "Ideal and non-ideal theories: the challenges of decision-making in an imperfect world" (Luna 2021).

F. Luna, *The Ideal and Non-Ideal in Bioethics*, The International Library of Bioethics 115, https://doi.org/10.1007/978-3-032-14594-9_2

our theoretical propositions to the complexities of everyday life. For those working in bioethics, this is perhaps the greatest challenge. Traditionally, bioethics has relied on practical yet often naïve theoretical frameworks—such as unexamined principles or values. Now, with the benefit of hindsight, the time is ripe to reassess the relationship between the ideal and the 'real' and to offer an analysis situated within the realm of meta-bioethics.

As we have seen, while at times, ITs or a partial IT seem to be the right answer, they can also fail to address the imminent challenges posed by our daily lives. Sometimes it can be the second-best or morally suboptimal alternative that provides the best answer. However, even the latter may not only leave moral residues and emotional after-effects (Tessman 2010), but may also prevent better options from emerging. Thus, selecting implementable options is rarely straightforward.

While the terminology of 'more or less idealised' theories aligns with public policy, where purely ideal theories are absent, it also fits with the endorsement of a continuum between ideal and non-ideal theories. Taking seriously into account this range of theoretical possibilities, we should ask ourselves several questions. What is the justification for the use of NITs, and when should we opt for ITs? We have just seen the decision-making mechanism at work, but why do we need such a theoretical apparatus? Are there valid uses or justifications for ITs and NITs?

This chapter strives to answer these questions. First, some distinctions are introduced, since most of the discussions are drawn from recent political philosophy[1] and are not so well known in the world of bioethics. Then we will see how ITs and NITs relate to each other, and how the initial and standard model of ITs has been challenged with the rise of NITs.[2] Redeemable and valuable elements of ITs should not be forgotten. We shoud not fall into the extreme of rejecting everything that is ideal, but rather, carefully assess which elements are harmful and should be avoided, and which should remain and be rescued.

The chapter concludes with one concept oriented towards the ideal and two that immerse us in the non-ideal: dynamic duties, vulnerability and autonomy. These ideas—dynamic duties, vulnerability, and autonomy—complement the preceding analysis. The perspective offered by this book differs from traditional accounts in revealing the connections each concept maintains with both the ideal and the non-ideal. To explore the path towards a more ideal world, Gilabert's proposal of dynamic duties is introduced. The concepts of vulnerability and relational autonomy are also presented, both of which establish strong ties to the non-ideal.

---

[1] For example, Simmons notes that there is very little literature on non-ideal theories (Simmons 2010: 1–2). Although there has been significant growth in political philosophy in recent years, these theorisations are little known and there is even less material applied to bioethics.

[2] Non-ideal theories driven, for example, by feminism, have begun to be re-evaluated in recent years. See Victor and Guidry-Gimes (2021).

2.2 A Continuum of Theories and the Decision-Making Mechanism

In this chapter I seek to avoid aligning myself exclusively with either of the usual traditions—those who defend the primacy of the ideal, and those who advocate for more realistic or non-idealised perspectives. There is often a tendency to assume that only one approach is valid and to dismiss the other entirely. Proponents of the ideal emphasise the importance of clear values and objectives, arguing that ideals should guide our actions. This, they claim, is what ethics is about—principles and values. They construct theories grounded in ideals, values and abstract principles, often dismissing the complexities of everyday life, the barriers and the setbacks that hinder progress towards the ideal. In their view, ideal answers are the ones we should pursue in order to create a better world. In contrast, those working on the front lines of reality—clinicians, researchers, policy-makers—are acutely aware of the myriad problems embedded in specific socioeconomic contexts. To them, such idealistic proposals may appear entirely unworkable. They navigate a landscape full of practical constraints that cannot simply be ignored, and the noble, longed-for ideals can seem entirely out of reach. Yet, in their critique of 'ivory-tower" idealism, they can become entrenched in the opposite position, denying any space for ideals altogether.

The answer, however, does not always lie at either extreme. Obstacles must be recognised, and new, non-ideal responses must be developed. But in other circumstances, it is equally necessary to look beyond the limitations of the *status quo* and to aim for more fundamental change—in short, to appeal to the ideal. As far as I am aware, this position has not been clearly articulated in the bioethics literature, where the debate is typically framed in binary terms: either a strong defence of the ideal or a commitment to the non-ideal. The previous chapter illustrated this point by showing how in practice there is no one theoretical proposal, be it an IT or an NIT, that should unfailingly guide us in the multiplicity of cases and policies that arise in the real world. So I argue that we must follow the winding and steep path that allows for the coexistence of both theoretical positions. To do so, I address a little-explored field.

How do we decide which of the ideal, non-ideal or second-best proposals we should follow? A fundamental issue in bioethics is that the choice of either an ideal or a non-ideal model cannot be arbitrary and must be made according to a certain logic and coherence. In other words, we must avoid being arbitrary when making a decision on how to solve certain problems. When we are considering a plurality of eligible theories (be these ideal or NITs), a decision mechanism with specific requirements is needed. The five decision-making criteria proposed as a first step in Chap. 1 help to provide coherence by taking into account the real world, the possibilities for action, as well as some normative elements when making decisions about which theory or type of theory to follow when confronted with concrete real-world problems. Thus, what was presented without further conceptualisation in the previous chapter plays a pivotal role when we endorse this vision of a plurality of eligible theories (be these ideal or NITs). The mechanism (illustrated in Fig. 1.1) provides a guide for clear deliberation, seeks to achieve consistency and at the same time avoids arbitrariness. If we are to provide an adequate response from bioethics, the five decision-making requirements should not be approached in a static way, or as an algorithm to be applied without reflection. We must consider the non-ideal world with its patterns of oppression, injustice, partial compliance, and so on. We must also take into account that actions or proposals that are currently not possible may be possible in the future. So as we will see, we must use a dynamic approach. We should use our political imagination and dynamic obligations or duties, as well as a reflective equilibrium, to select the best proposal.

2.3 Conceptual Distinctions

Let us begin with the ideal and non-ideal problem. In recent years, some scholars from the field of political philosophy have questioned the methodology for elaborating normative prescriptions (Mills 2005; Sen 2006). Much work in political philosophy (formulations about justice, or how to make equitable distributions) was considered flawed because it was of little or no practical help. This realisation pointed the way to a methodological shift (Valentini 2012) from John Rawls (1971) as the paradigmatic model to the search for other alternatives, including NITs. Much of the IT and NIT debate, at least initially, seemed to focus on Rawls as the target of many of these analyses (see Valentini 2012). This book is not one of exegesis so I will consider Rawls's position only when it is relevant to, or illustrates a point of, my argument.

The analysis and discussion of ITs and NITs is very broad and encompasses numerous conceptual issues and distinctions that will not be presented in this chapter. Instead, I aim to clarify what ITs and NITs are, because although this is not as simple as it might sound, it will help to clarify the field of study. There are several ways to consider what it is, as well as to identify different proposals. I will present Laura Valentini's (2012) formulation as it offers a clear view of some of the issues at stake.

Valentini explains that, in a first sense, IT can be seen as a theory of *full compliance*, i.e., rules are universally enforced and no one is left out, and NIT can be understood as a theory of *partial compliance,* where only some people comply with their duties and obligations. Here an important debate is the question of what duties and obligations are incumbent upon us in situations of partial compliance as opposed to situations of full compliance. For example, you may ask: should people who fulfil their obligations be responsible for what non-fulfillers fail to do? If so, how?

Secondly, IT can be understood as a *utopian* or idealistic theory and NIT as a *realistic* theory. In this second reading, the debate centres on the question of whether feasibility considerations[3] should constrain normative political theorising and, if so, what kind of feasibility constraints should matter. Another debate examines the degree of

[3] Regarding feasibility concerns, Gilabert (2017) offers a systematic approach to the relationship between feasibility and justice. See also Galston (2010), who proposes a reconstruction of realism.

idealism or realism required. For Gerald Cohen's (2003, 2008) utopian position, for example, principles of justice must be entirely independent of the constraints of the real, as opposed to the more realist analysis presented by Rawls. For Rawls, principles of justice are developed in response to the problems of the human condition, moderate scarcity and limited altruism (Rawls 1999: 1922). However, there are other realist positions, such as that of Waldron (1999), which considers that Rawls was excessively idealistic and failed to consider reasonable disagreement about justice.[4]

The third way of understanding IT indicates that it is an *end-state* theory and that NIT can be thought of as a *transitional* theory. This debate centres on whether a normative political theory should aim to identify an ideal of social perfection or whether it should focus on transitional improvements without the need to determine what the optimal situation is. For example, while IT sets a long-term goal for institutional reform, NIT asks how this long-term goal could be achieved, or worked towards, usually in incremental steps (Valentini 2012; Laurence 2021).

We can infer that each understanding of IT and NIT has different objectives and scopes. It is important to know that these differences exist because they help us to understand the different ways in which we can interpret and apply an IT or a NIT. Moreover, there are some positions that tend to blur the categorical distinction between ITs and NITs and speak of a "territory of a multidimensional continuum". For example, Hamlin and Stemplowska maintain this position, although they distinguish four broad approaches:

(1) total and non-total compliance;
(2) idealisation and abstraction;
(3) fact-sensitivity and fact-insensitivity; and
(4) perfect justice and local enhancement of justice.

They make a categorical distinction between the theory of ideals (identification and elucidation of ideals) and the continuum of ITs and NITs that are identified with institutional designs (Hamlin and Stemplowska 2012).[5] If we bear Valentini's distinction in mind, then we must also recognise that for her, these are not radically different areas and that some of her debates are interrelated. This chapter focus primarily on some of the issues that correspond to what Valentini sees as realist and transitional NITs, that is, the second and third meanings.

---

[4] Other political realists, for example, include power structures (Williams 2005).

[5] For details see Hamlin and Stemplowska, 2012: 48–49.

2.4 Ideal Versus Non-Ideal: Sen's Criticisms

This chapter starts from the assumption that the pendulum has now swung in favour of NITs. There is scant literature in bioethics on the non-ideal. However, it has been included in some recent feminist proposals (Victor and Guidry Grimes 2021).[6] The feminist viewpoint takes the context very much into account and subtly perceives the various oppressions to which women are subjected. It is therefore logical that feminists were among the first in the bioethics world to adopt a questioning view of the ideal in bioethics.

If we accept the richness, subtlety and complexity of the non-ideal challenges, we should ask ourselves why the ideal has lost ground and whether there is anything to rescue from these traditional ideal models.

Going back through history, ethics and political philosophy have generally been handled with a high degree of abstraction. They tell us what we should do, what is right or good, or how a society should be ordered in order to be fair. But why then have ITs been subject to such criticism? Precisely because of their abstraction from real conditions, for their lack of consideration of oppressive and distorting situations, and for their use of certain over-idealised concepts that hide relevant features when making moral assessments.

One of the classical critics is Amartya Sen (2009). Sen criticises transcendental approaches—mostly represented by Rawls—as opposed to the comparative approaches he advocates.[7,8] In his questioning of Rawls, Sen (2009) explains that transcendental propositions are not feasible, and that there may be no reasoned agreement on the nature of the just society. He criticises the possibility of unanimous agreement on principles with a complete ordering. However, notwithstanding Sen's attacks, in this debate Rawls does not seem to be quite so prescriptive. He argues: "... until the ideal is identified, at least in outline ... nonideal theory lacks an objective, an aim, by reference to which its queries can be answered" (Rawls 1999, 90). Rawls does not posit the need for a full description or identification of this ideal,[9] but makes a priority claim that entails other commitments.[10] Nevertheless, I agree with Sen's main point: we do not need a finished description of the perfect society or a fully ordered set of principles agreed upon by all. However, I also concur with Rawls in that IT is needed at least in a broad outline since it plays a relevant role (as I will argue

[6] I took part in this compilation with an early version of the reflections presented in this book (Luna 2021).

[7] I will use the terms 'ideal theory' (IT) and 'transcendental approach; (TA) interchangeably, as well as 'non-ideal theory' (NIT) and 'comparative approach' (CA). Robeyns (2008) also considers Sen's terminology similar to IT and NIT.

[8] I do not attempt an exegesis of Rawls' position nor of Sen's criticisms in this regard.

[9] On this point I follow Gilabert. This author explains that Rawls never says that he has demonstrated that his three principles are better than all conceivable alternatives (Gilabert 2008, See note 4, page 8).

[10] The question of whether Sen's analysis of Rawls is correct or not is beyond the scope of this book.

below).[11] In Pablo Gilabert's words: political philosophy should not be ashamed to pursue ambitious theories of justice, but it must also engage in careful considerations of feasibility and viability in relation to their practical implementation (2012: 39). And I would add that it is bioethicists who can collaborate towards the latter task with a strong understanding of the empirical challenges.

However, this does not imply advocating a *priority claim* to the ideal. The design of principles does not require presupposing a controversial claim to the existence of a timeless and independent moral truth. The method presented in Chap. 1 attempts to systematise our considered judgements in a reflective equilibrium:

> The method of reflective equilibrium consists in working back and forth among our considered judgments (some say our "intuitions") about particular instances or cases, the principles or rules that we believe govern them, and the theoretical considerations that we believe bear on accepting these considered judgments, principles, or rules revising any of these elements whenever necessary in order to achieve an acceptable coherence among them. (Daniels 2011)[12]

Daniels argues for a broad reflective equilibrium that includes three sets of elements: background theories, considered judgements or intuitions, and principles to avoid moral intuitionism or subjectivism.[13] I consider reflective equilibrium to be highly relevant to bioethics as we work with and evaluate real-world cases and situations that may challenge our intuitions, underlying theories, conceptual tools or basic principles. In this framework—as we will see—an idealised theory or a set of idealised principles may still be relevant to our political, practical and strategic thinking.

The second criticism Sen considers is that of redundancy. Even if a perfect society could be identified, it would be of no use in benchmarking, as it is neither necessary nor sufficient, and the transcendental approach is not productive. While Sen is right that such an approach may not be sufficient, no reasonable advocate of it would argue that it alone offers a solution to all comparative questions (Gilabert 2012: 10). However, the accusation of being or not being useful seems too strong.

In this respect it is worth considering the reflections of Gilabert. He provides several arguments against Sen's assertions. I focus on three that I believe can help us in our further research:

(1) Gilabert defends the *heuristic importance* of exemplars and explains that even if we do not know exactly how to judge the comparative distance between one alternative and another from the perfect exemplar, the enquiry into how the alternatives relate to the transcendental approach will shape our comparative enquiry in an interesting and relevant way. We could, in principle, propose a partial ordering or discard some options.

---

[11] This is also the position of Robeyns (2008: 344).

[12] Reflective equilibrium is endorsed by several theorists, such as Valentini (2012: 6), and Gilabert (2012, 2017).

[13] There is also narrow reflective equilibrium that would not include the sets of background theories (Daniels 1979: 257–259).

(2) Another very interesting argument is that a transcendental approach offers a way to *critique the status quo*. Gilabert argues that a theory of justice should help us to identify processes of improving justice in both the short and long term. Focusing only on manifest or overt injustice can lead to a failure to deepen research to identify important forms of injustice not yet perceived, and lead us to overlook the fact that the feasibility of political action is open to temporal variation: what is not feasible now may be feasible in the future, provided we take the right steps. The pursuit of just societies would keep us open to expanding our epistemic and practical understanding of an improvement of justice, so as not to surrender to injustices and allow us to fight back (Gilabert 2012: 15). Although Gilabert does not adhere to the need for a perfectly just society, he refers to it (p.15, ibid.) and points out that research into it is important, if not indispensable, if we are to make comparative ethical assessments. I myself agree with a slightly weaker version gives prominence to the ideal without succumbing to it. That is, we should not always give priority to the ideal, at least without justification.
(3) Finally, the third argument discusses the importance of *inspiration and motivation*. Having an ideal of society can motivate or inspire us to make changes.

While Gilabert acknowledges that his arguments fall short of demonstrating that it is strictly necessary to imagine a transcendental approach, they do show that the transcendental approach is not irrelevant and can contribute to the normative elaboration of comparative approaches to feasible alternatives (2012: 17). I consider this sufficient for our purpose. We do not need stronger commitments.

2.5 Ideal Versus Non-Ideal: Mills's Arguments

The philosopher Charles Mills (2005) argues that bad idealisations should be avoided. I wish to make it clear that I agree with most of the criticisms of Mills, who, on the contrary, rejects the ideals proposals. Mills points to the failure on the part of mainstream ethics and political philosophy to recognise structures of oppression and exclusion. Mills explains that the IT approach uses some or all of the following concepts and assumptions as its basic apparatus: an idealised social ontology (characterisation of human beings); idealised capacities (completely unrealistic capacities attributed); silence about oppression (historic oppression and its legacy in the present or current ongoing oppression); idealised social institutions (the family, economic structures, the legal system); an idealised cognitive sphere (general social transparency will be presumed, with cognition limited to self-interest bias); and strict compliance (Mills 2005: 168–169). As we shall see in Part III of this book, all of these criticisms are applicable to the traditional concept of autonomy that feminism reconceptualises.[14] Even if we recognise the existence of these assumptions in most ideal theories, I do not think we should reject a priori all ITs or idealised propositions (in the previous chapter we also saw how some of these assumptions creep into non-ideal propositions), as certain features of ITs remain relevant. I argue that both ITs and NITs can have a place but that this must be adequately justified. I believe that Mills rejects over-reliance on ITs because he has in mind the prevailing ethical and political philosophy of the late twentieth century, in which the assumptions of ITs were uncritically accepted and their principles took absolute priority.

Moreover, I believe that Mills, while decrying ITs, accepts a continuum between the different theoretical propositions. For example, he says: "Nor does the simple appeal to an ideal (say, the picture of an ideally just society) necessarily make the theory ideal theory, since nonideal theory can and does appeal to an ideal also" (Mills 2005: 171). Thus, at least tacitly, Mills accepts different levels of idealisation in NITs, in addition to the fact that NITs can use ideals. Thus, there are no 'pure' NITs, but they also include different idealisations and assumptions.

For Mills (and for myself) it is imporant to adopt a critical attitude towards the ideals, principles and conceptual tools we accept and the consequences they may have. This aligns well with reflective equilibrium, which fosters dynamism and critical tool assessment. This process allows us to analyse and filter out bad idealisations or problematic principles.

Thus, this proposal argues for a pluralistic view in which we can find more idealised visions, ITs or transcendental approaches that may be useful on certain occasions, and NITs or comparative approaches that are applicable in other cases. Some NITs can be critical and provide a good empirical relationship to conditions of oppression, for example, by offering better prescriptions to guide action. Other NITs may provide rectificatory justice (Pateman and Mills 2007: 94f). IT should not always prevail, nor should it always be the second-best or realistic alternative.

The opposition between IT and NITs seems to be artificial. Philosopher Ingrid Robeyns also argues for a continuum from IT (as the ultimate goal) to NITs. She states that "Ideal and nonideal theories of justice tell us what ideals we are striving for, how different principles of justice should be weighed against each other, how justice needs to be balanced against other values, and how to deal with instances of widespread noncompliance" (Robeyns 2008: 349). In a similar vein, theorists Hamlin and Stemplowska speak of continuous variables: "Compliance, idealisation, abstraction, fact-sensitivity and improved realisation of a value are all matters of degree and of 'appropriateness' … any sharp categorical distinction between ideal and non-ideal theories that focuses on a single dimension seems implausible at best" (Hamlin and Stemplowska 2012 49). I concur with this view. We should also seek to avoid dichotomies and reductionism in which only one alternative (IT or NIT) is valid.

---

[14] For example, the traditional conception of autonomy assumes that people function in isolation, with a strong emphasis on rationality, and that different forms of oppression such as gender have no place within its analysis. This will be made particularly explicit in the final chapter on relational autonomy.

Finally, I agree with Gilabert that the real question is not whether we should select NIT instead of IT, but rather how we can select both in an integrated way (Gilabert 2012: footnotes 17, 22). To do so we need to include some criteria that help in decision-making, as we saw in the previous chapter. Now we can also understand how this mechanism plays out within a pluralistic view on ideality.

2.6 Good and Bad Idealisations

While some theorists recognise the usefulness of ITs or more idealised theories, there remain strong objections not only to IT but also to the use of idealisations. Ingrid Robeyns explains that idealisations by definition are simplifications and often represent a much better reality than what actually exists. Most or even all ITs employ idealisations. One of the functions of such idealisations is to keep the complexity of the theory within manageable limits (as it reduces the number of parameters the theory has to deal with). According to Robeyns we can ignore certain injustices and their causes because at the ideal level these injustices simply should not occur (2008: 354). In her examination of how to distinguish between good and bad idealisations, Robeyns states that in the literature IT is often confused with theories based on idealising assumptions and, in particular, with idealisations that are bad idealisations. She argues that "the role of ideal theory is limited, perhaps more so than generally acknowledged in the literature, and that these limitations should be much more explicitly discussed by ideal theorists" (2008: 342). According to Robeyns, IT plays a role, but assumptions of bad idealisations should be filtered out, and scholars need to correct the current academic bias towards ideal theory and do much more theoretical work to bridge the gap between the ideal and the non-ideal (Robeyns 2008: 342). I fully agree with these statements.

Why is there so much resistance to idealisations? In addition to an allergy to counterfactual thinking and highly abstract reasoning. Robeyns points out that the transition from IT to NIT is anything but straightforward. Thus, the more an IT has been built on idealisations, the further it is from offering a clear guide to the non-ideal world. She argues that the problem is not that idealisations are not acceptable at the ideal level. The problem is how to deal with idealisations when we move to the non-ideal level.

Another function of idealisations lies in the purpose of IT itself: to model the desirable properties of the ideally just society (e.g., we would not want people to act on prejudices and stereotypes).[15] However, bad idealisations cannot be justified theoretically. They ignore certain forms of injustice that need to be theorised (aspects that are relevant for some groups in society more than others). One example is the idealisation that assumes that people are totally independent instead of including within the notion of personhood the need to receive and provide care: "The fact of human dependence on the care of others is like the fact of human mortality: both

[15] Robeyns explains that Dworkin uses this strategy (Robeyns 2008: 354).

are facts that no serious comprehensive theory can ignore" (Robeyns 2008: 358). Thus, Robeyns rightly argues that this idealisation favours those who benefit from the current unjust arrangements. I also agree with one of Mills' main criticisms, namely, the contribution to the perpetuation of illicit privileges in certain groups (Mills 2005: 166). Robeyns explains that care is not equally distributed among all people, with women providing much more care than men (both paid and unpaid), and that caregiving is overrepresented among female caregivers, immigrants, low-income women, and women of colour (Robeyns 2008: 359). This creates an ideological bias in the theory: bad idealisations imply biases towards groups that are treated unfairly, even if the theory-maker did not intend this. Similarly, some ideal social institutions (the family, the economic structure, the legal system, etc.) can be deeply problematic, and such idealisations should be avoided when they increase the privilege of some already privileged groups (Mills 2005: 169). 'Christian List and Laura Valentini also ask whether a theory contains the 'right' idealisations in relation to its purpose, but propose a different analysis. They distinguish between a theory that is abstract, i.e., says nothing about an issue, and a theory that is idealised, i.e., implies some simplifying or limiting falsehood about that issue (List and Valentini 2016: 19–20).[16] Moreover, it is worth specifying that some degree of idealisation can play an important and justified heuristic role (O'Neill 1996, Chap. 2; Farrelly 2007). List and Valentini distinguish three possible levels (loci) at which idealisations can occur and assess some of them as more problematic than others. These loci are:

(i) the theory itself;
(ii) the conditions of application of the theory's prescriptions (i.e., "if such and such conditions are given, then such and such situations follow"); and
(iii) the justification of the theory.

List and Valentini consider the first level to be the most problematic. The other two may make theories less relevant to the real world because their prescriptions may not be valid in the second case and will not be sufficiently justified in the third. However, they do not present the threat that the first level possesses (List and Valentini: 20–21).

These authors provide two examples from Rawls' (1971) *Theory of Justice.* First, they examine Rawls' assumption of "a society existing under favourable historical and social conditions" and consider that this idealisation operates at the level of the condition of application of the theory's prescription, i.e., at the second level. Second, they argue that the often criticised 'idealised' assumption of 'full compliance' is not made within the theory itself, but occurs as part of the justification Rawls offers for the theory. This is one of the assumptions made by supporters of the thought experiment of the original position. It is therefore a third-level idealisation.

Thus even those who are strongly inclined towards non-ideality accept the value of idealisations. Even Mills, in a somewhat extreme characterisation of IT, assumes that the use of ideals is permissible:

[16] I will not enter here into the debate between idealisation and abstraction that goes back to O'Neill (1987).

> Now what distinguishes ideal theory is not merely the use of ideals, since obviously nonideal theory can and will use ideals also (certainly it will appeal to the moral ideals, *if it may be more dubious about the value of invoking idealized human capacities)*. What distinguishes ideal theory is the reliance on idealization to the exclusion, or at least minimization, of the actual. (Mills 2005: 168, my emphasis)

Lisa Tessman, who also supports the work and value of NITs (2010: 810), argues for the presence of ideals in NITs. Most interestingly, she distinguishes between two types of ideals. The first is one that guides action and is achievable, i.e., feasible ideals. The second type is unattainable but worthy. Tessman identifies a shifting line between what is attainable and what is unattainable. She recognises the value of differentiating options according to degrees of probability, that is, according to their possibility/impossibility and according to whether they are achievable individually or only achievable if there is a collective of cooperating agents. Unattainable but worthy ideals can be included after critical examination. Tessman criticises NITs that focus exclusively on issuing guidance for appropriate action and calls for truly worthy ideals to be accommodated. Thus, while bad idealisations are problematic and should be avoided, this does not mean that all idealisations or worthy ideals should be discarded.

2.7 Different Perspectives on IT

So far we have seen that ITs do not necessarily have the paradigmatic role they traditionally enjoyed, and need not necessarily be the focus of theorisations that structure every implementation in the real world. Let us reconsider what the role of IT might be under more meticulous and critical examination. We may have stronger or weaker versions of the value or usefulness of IT. For example, for John Simmons, unless we have an IT that states.

i) what counts as permissible and
ii) what counts as successful (i.e., full justice/approaching full justice),

We cannot establish whether or not our transitional recommendations meet these requirements (Simmons 2010: 34). This is a rather strong position that implies some prioritisation of the ideal. Philosopher Ingrid Robeyns provides another version: for her, IT plays a limited role and offers a rather graphic image of a mythical Paradise Island we would all love to inhabit, but no one tells us how to get there, nor whether it is even possible, since no one has ever set foot on such an island. However, our ultimate goal of reaching Paradise Island tells us the direction in which we must move (Robeyns 2008: 345). Robeyns clarifies that the NIT will bring us closer to the island, insofar as it leads us towards the better design and implementation of justice, providing the theoretical foundations for figuring out what to do to get us closer to that perfectly just society (Robeyns 2008: 345).

Robeyns characterises NITs as having two main functions:

(1) to compare between different social states to enable us to make and assess which is more just, and
(2) to guide our actions towards the ideals of such a society.

However, she acknowledges that the second function may be more difficult to achieve, given the gap between ITs and NITs (ibid., 355–358). Robeyns criticises Rawls' conceptualisation of NITs. She does not accept that war or just punishment are relevant for an NIT and claims that not enough emphasis is placed on some of the issues of NITs, such as how to weigh different principles of justice and whether such principles should be adapted (ibid., 346–347).[17]

There are still weaker versions of IT, such as Pablo Gilabert's arguments against the *redundancy charge*. Gilabert speaks of the heuristic importance of exemplars—their usefulness in the partial ordering or discarding of alternatives. He also argues that it presents a way of criticising the *status quo*. In line with Robeyns and Simmons, Gilabert also argues that IT helps to identify long-term processes for the improvement of justice. Finally, he argues that IT can be an inspiration and can be relevant with respect to motivation. In my view, we can work with this weak proposition, sustain a pluralistic approach and consider a continuum from IT to NITs.

The final set of questions "we must ask ourselves" comprises: is it better to stay on the nearby sub-optimal island or should we try to continue, through stormy and uncertain waters, to Paradise Island? What should the decision-making process look like? And if we can't set foot on Paradise Island, can we at least try to take some pictures, draw maps, plot possible routes and then return better equipped to do the work of making the mud of real life more paradisiacal? Can we follow specific priority rules, algorithms or procedures? Or are these kinds of theoretical artefacts useless, too narrow, limiting or restricting thinking?

2.8 Other Proposals: Dynamic Duties

At this point, we are confronted with the possibility of theoretically rescuing or defending certain ITs that enable us to consider the ideal, while also recognising a spectrum of options among NITs or secondary alternatives. However, the non-ideal world challenges not only theories but also various key concepts. We now reconsider some 'traditional' concepts and their implications—which we shall be illustrating in the final chapters of this book—through the lenses of the ideal and the non-ideal.

Gilabert's proposal contains a particularly interesting idea worth bearing in mind as we reason. He presents an interesting dynamic approach to the relationship between justice and feasibility, which he applies to what he regards as the three dimensions of justice. He explains that normative desirability and feasibility interact at different levels of deliberation on the pursuit of social justice, identifying demands that are both normatively desirable and normatively feasible for each of the dimensions of justice. These dimensions encompass normative demands on:

[17] See Amanda Roth (2021) for further discussion.

DI: Core principles, which include evaluative and prescriptive principles;
DII: Institutions and social practices;
DIII: Processes of transformation.[18]

Gilabert argues that we can maintain and combine idealism about principles and realism about feasibility (Gilabert 2017: 95–96). He introduces the innovative idea of *dynamic duties*. He argues that when demanding principles collide with contingent constraints, an appropriate response is to use our *political imagination* to envision alternative ways of fulfilling principles in different contexts and to recognise *dynamic duties* to expand our ability to fulfil these principles over time. For Gilabert, dynamic duties—unlike 'normal' duties—aim not at outcomes within current constraints but at changing those constraints. They aim to change the circumstances so that desirable outcomes are achievable (Gilabert 2017: 119). Dynamic duties direct a change, often an expansion, of an agent's power to achieve certain outcomes.

He also proposes a deliberative, reflective equilibrium as the methodology to adopt. For him there is a logical sequence in the overall conception of justice (the final product has to exhibit a logical sequence) and there is an epistemic back and forth that determines what to accept as the content of each dimension of justice. Gilabert argues that his interpretation differs from that of Rawls.[19]

Note that the first two requirements of our proposed five-requirement decision-making mechanism are directly related to feasibility, taking into account the constraints and challenges of the real world, while the three remaining requirements point to moral content, justice and long-term aims.[20] This mechanism should enable us to make decisions as to which theoretical proposition to accept and prevent such choices from being arbitrary. We must also bear in mind that actions or proposals that are currently not possible may be possible in the future. So as stated above, we must use a dynamic approach as we scan the horizon of ITs and NITs, taking care not to do it in a static way or as an algorithm to be applied without reflection. In addition, we should use our political imagination and dynamic obligations or duties, as well as a reflective equilibrium, to select the best proposal. Equipped with this conceptual framework we must consider the non-ideal world with its patterns of oppression, injustice, partial compliance, and so on, and seek the best answer, be it more ideal or non-ideal.

[18] See Gilabert 2017, 106.

[19] Gilabert describes three dimensions of justice (basic principles/ institutions and social practices/ processes of transformation), arguing for a progression from more abstract to more specific prescriptive judgements. And it is this epistemic back and forth that determines which moral orientation to accept (Gilabert 2017: 110–111, note 23).

[20] See the first chapter for details, or Figure 1.

2.9 Other Proposals: Revisited Concepts in a Non-Ideal Key

Just as the concept of dynamic duties draws us closer to the ideal world and encourages us to consider new possibilities for fulfilling principles, other concepts traditionally analysed in bioethics immerse us in non-ideality. Being aware that different types of concepts can be interpreted through both ideal and non-ideal lenses—not only theories framed as ideal or non-ideal—enriches our understanding of bioethics. Two such key concepts are *vulnerability* and *autonomy.*

The concept of vulnerability has traditionally been addressed without adequately considering context—that is, without sufficiently accounting for the non-ideal conditions inherently linked to the term. It is important to note that multiple vulnerabilities are closely connected to socioeconomic contexts, as well as to the legal and institutional frameworks of each country that impact those considered 'vulnerable'. Consequently, the non-ideal perspective moves away from a minimal conception of vulnerability—often equated simply with a person's competence—to a dynamic and far more flexible analysis.

In contrast, the traditional view of vulnerable populations offers an essentialist response that labels and stigmatises entire groups. It also rests on certain idealisations and assumptions that are problematic—for instance, in research ethics, the 'ideal' participant is often assumed to be an adult, minimally literate, capable of understanding, etc., while those who deviate from this ideal model are deemed 'vulnerable'. Once the relevance of context and non-ideality is acknowledged, the theoretical approach must be revised. The notion of 'vulnerable populations' should be abandoned. I propose, instead, a layered approach which is closely related to the non-ideal analysis. This complex strategy is revisited and made explicit in the final chapters of the book.

Another important concept, to asses and incorporate into the analysis is autonomy. This concept is autonomy, is central to bioethics, arguably emerging alongside it by placing patients' wishes at the forefront and challenging the traditional paternalistic model of medicine. While autonomy at first glance may seem unrelated to non-ideality, feminist critiques expose the idealisations embedded in the traditional view (i.e., an idealised view of human agency that is based on free and rational agents). In this traditional conception we can again see an idealised reading of this concept, a conception of a person that does not exist. In the last chapter, for example, I introduce Fineman's "myth of autonomy" as well as the impact it can have in the organisation of a society. As a way out of bad idealisations and problems related to this concept, some feminists offer new conceptualisations. One such conceptualisation is *relational autonomy*, which takes context very seriously. It moves away from thinking about individuals as independent, isolated beings and instead considers how context, family, and society influence people's 'autonomous' decisions.

A final point to note is that only by reconceptualising these ideas through a non-ideal lens does their relationship shift. Rather than functioning as opposites or counterpoint—where those who are not autonomous are simply vulnerable, implying a minimal view of vulnerability—these concepts can be seen as complementary. The strategy is to reveal the non-ideality also present within these terms. Incorporating the non-ideal lens into our analysis of concepts such as vulnerability and relational autonomy leads to theoretically relevant propositions. Moreover, and perhaps more importantly, analysing these concepts from a non-ideal perspective has substantial practical significance.

2.10 Conclusion

This chapter has argued for a rethinking of bioethics and its operation by considering how ideal and non-ideal theories relate to the real world. Further, it has sought to clarify the relationship between the ideal, the non-ideal and bioethics, as well as to examine some of the challenges we face when engaging with the complexities of the real world. It involves a novel meta-analysis of bioethics itself. However, it remains exploratory in nature.

Together with the decision-making mechanism, this chapter provides the theoretical framework of the book. It makes explicit the assumptions and elements that will be used to analyse concrete situations in the forthcoming chapters. The chapter advocates for a coherentist methodology grounded in reflective equilibrium. As noted, this reflective balancing act operates through the interplay between theoretical perspectives and practical cases. There is no single type of theory—ideal or otherwise—that can guide all real-world cases or policies. The reality is more complex: we must carefully assess the implications of these theories and apply the five criteria of the decision-making mechanism. We need to examine how various forms of idealisation function, and identify which inadequate forms should be discarded. Finally, we must analyse how both ideal and non-ideal theories interact with the *status quo*.

The next two chapters employ this conceptual framework to evaluate diverse responses to the COVID-19 pandemic. As bioethicists, our aim is to make the world a better place, and the concept of dynamic duties assists in promoting change and progress. In Chap. 5 we use this idea to explore ways of shaping a better future.

Thus, when practising bioethics—whether addressing a case or advocating for guidelines or policies—we must employ multiple tools, including the decision-making mechanism, to consistently choose between idealised and non-ideal proposals. Finally, two concepts that can be interpreted from a non-ideal perspective—vulnerability and relational autonomy—are introduced. These will be addressed in detail in the third part of the book.

References

Cohen, G. A. 2003. Facts and principles. *Philosophy & Public Affairs* 31 (3): 211–245. https://doi.org/doi.org/10.1111/j.1088-4963.2003.00211.x.

Cohen, G.A. 2008. *Rescuing justice and equality*. Cambridge: Harvard University Press. *England Journal of Medicine* 337 (12): 853–856.
Daniels, N. 1979. Wide reflective equilibrium and theory acceptance in ethics. *The Journal of Philosophy* 76 (5): 256–282.
Daniels, N. 2011. Reflective equilibrium. *The stanford encyclopedia of philosophy*, https://plato.stanford.edu/archives/spr2011/entries/reflective-equilibrium/
Erman, E., and N. Möller. 2019. A world of possibilities: The place of feasibility in political theory. *Res Publica* 1–23. https://doi.org/10.1007/s11158-018-09415-y.
Farrelly, C. 2007. Justice in ideal theory: A refutation. *Political Studies* 55 (4): 844–864.
Galston, W. A. 2010. Realism in political theory. *European Journal of Political Theory* 9 (4): 385–411.
Gilabert, P. 2008. Global justice and poverty relief in nonideal circumstances. *Social Theory and Practice* 34 (3): 411–438.
Gilabert, P. 2012. Comparative assessments of justice, feasibility, and ideal theory. *Ethical Theory and Moral Practice* 15 (1): 39–56.
Gilabert, P. 2017. Justice and feasibility. A dynamic approach. In *Political utopias. Contemporary debates*, eds. Michael Weber and Kevin Vallier, 95–126. Oxford: Oxford University Press.
Hamlin, A., and Z. Stemplowska. 2012. Theory, ideal theory and the theory of ideals. *Political Studies Review* 10 (1): 48–62. https://doi.org/doi.org/10.1111/j.1478-9302.2011.00244.x.
Laurence, B. 2021. *Agents of change*. Cambridge (USA): Harvard University Press.
List, C., and Valentini, L. 2016. The methodology of political theory. In *Oxford handbook of philosophical methodology*, ed. Herman Cappelen, Szabó Gendler, Tamar, and Hawthorne, John, 525–553. Oxford: Oxford University Press. https://doi.org/10.1093/oxfordhb/9780199668779.013.10
Luna, F. 2015. Medical ethics and more: Ideal theories, non-ideal theories and conscientious objection. *Journal of Medical Ethics* 41: 129–133. https://doi.org/10.1136/medethics-2014-102295.
Luna, F. 2021. Ideal and non-ideal theories: The challenges of decision-making in an imperfect world. In *Applying nonideal theory to bioethics: Living and dying in a nonideal world*, ed. Victor, E. and L. Guidry-Grimes. Springer: Philosophy & Medicine Series No. 139, 17–40.
Mills, C. W. 2005. 'Ideal theory' as ideology. *Hypatia* 20 (3): 165–184.
O'Neill, O. 1987. Abstraction, idealization and ideology in ethics. In *Moral philosophy and contemporary problems*, ed. John David Gemmill Evans, 55–69. Cambridge: Cambridge University Press. https://doi.org/10.1017/s1358246100003660
O'Neill, O. 1996. *Towards justice and virtue: A constructive account of practical reasoning*. Cambridge: Cambridge University Press.
Pateman, C., and C. Mills. 2007. *The contract and domination*. London: Polity.
Rawls, J. 1971. *A theory of justice*. Cambridge: Havard University Press.
Rawls, J. 1999. *The law of peoples*. Cambridge: Harvard University Press.
Robeyns, I. 2008. Ideal theory in theory and practice. *Social Theory and Practice* 34 (3): 341–362. https://doi.org/10.5840/soctheorpract200834321.
Roth, A. 2021. Nonideal theory and ethical pragmatism in bioethics: Value conflicts in LGBTQ+ family-making. In *Applying Nonideal Theory to Bioethics: Living and Dying in a Nonideal World*, ed. Victor, E., and Guidry-Grimes. L.K, 375–396. Springer.
Sen, A. 2006. What do we want from a theory of justice? *Journal of Philosophy* 103 (5): 215–238.
Sen, A. 2009. *The idea of justice*. London: Allen Lane/Penguin.
Simmons, J. A. 2010. Ideal and nonideal theory. *Philosophy & Public Affairs* 38 (1): 5–36.
Tessman, L. 2010. Idealizing morality. *Hypatia* 25 (4): 797–824.
Valentini, L. 2012. Ideal vs. non-ideal theory: A conceptual map. *Philosophy Compass* 7 (9): 654–664.

Victor, E., and L. Guidry-Grimes, eds. 2021. *Applying nonideal theory to bioethics: Living and dying in a nonideal world*. Springer: Philosophy & Medicine Series No. 139.
Waldron, J. 1999. *Law and disagreement*. Oxford: Clarendon.
Williams, B. 2005. *In the beginning was the deed*. Princeton: Princeton University Press.

Part II
Global Justice in a Non-Ideal World

Chapter 3
Finding Answers to the Non-Ideal World of the Pandemic

Abstract The third chapter identifies some debates and ethical issues that arose in relation to the equitable distribution of vaccines during COVID 19 pandemic. It explains the Hobbesian state of nature in which countries were engaged in a kind of every-man-for-himself power struggle and the prisoner's dilemma the world faced during the pandemic. This chapter outlines some of the approaches to the so-called "vaccine nationalism", their justifications and the problems involved. A moderate nationalism, like the 'flu risk' criterion is explored. The chapter also examines these different strategies, and how much of the ideal approaches and the explicit criteria of the decision-making mechanism outlined in the first chapters can be applied, and in what ways.

Keywords Ideal and non-ideal theories · Decision-making mechanism · COVID 19 pandemic · Vaccine global allocation · Prisoner's dilemma · Vaccine nationalisms · Cosmopolitan view · Flu risk criterion

3.1 Introduction

The suffering caused by the COVID-19 pandemic was of unprecedented proportions. Its impact extended beyond health, affecting the social and economic lives of most people, and exacerbated existing fragilities and inequalities. Inequitable vaccine distribution, for example, was described by UN Secretary-General António Guterres as the greatest moral failure of our time (UN 2022).

When I wrote most of the content of this second part of the book, we were still living through the pandemic. The World Health Organization (WHO) first declared COVID-19 a public health emergency of international concern on 30 January 2020. It was declared a pandemic on 11 March 2020, until May 5th 2023. However, experts predict a 50% chance of another pandemic in the next 25 years (Schaefer et al. 2023).[1] Thus, while this chapter analyses the COVID-19 years, its reflections and proposals

[1] During 2022, smallpox appeared, but it is predicted that different viruses will continue to appear globally.

F. Luna, *The Ideal and Non-Ideal in Bioethics*, The International Library of Bioethics 115, https://doi.org/10.1007/978-3-032-14594-9_3

extend beyond them, and consider the contributions of bioethics to how we should face a new pandemic. In short: what went wrong, and how can we avoid repeating mistakes?

If future-oriented thinking is to be useful, while many empirical elements of the COVID-19 pandemic are likely to vary (the type of inputs needed, the populations with the greatest vulnerabilities, etc.), the global distribution of scarce health resources (medicines, vaccines or other materials) among unequal countries is likely to remain one of its challenges.

The aim of this chapter is to assess what responses were made in this non-ideal world, how they should have been acted upon, and how to think about a future pandemic. We will also examine how much of the ideal approaches and the explicit criteria of the decision-making mechanism outlined in Chapter 1 can be applied, and in what ways.

I begin by describing the pandemic years and how what happened can be analysed in terms of the prisoner's dilemma. I then outline some of the approaches to so-called 'vaccine nationalism', their justifications and the problems involved, evaluating them in the light of proposed ideal and non-ideal theories. Chapter 4 will consider the ethical plausibility of bilateral donations offered by some countries and analyse the multilateral COVAX mechanism as another option, and Chapter 5 considers possible alternatives in the form of *dynamic duties*. These dynamic obligations should also complement some of the proposals put forward and promote change to the injustices of the real world. And, even if we cannot yet set foot on the island paradise that the ideal represents, we may be able to face the next pandemic in a better state not only in empirical but also in ethical terms.

3.2 A Non-Ideal World

In early 2020, when no effective treatments for COVID-19 were available, vaccines were presented as one of the best solutions. From the start of the pandemic, attempts were being made to develop a vaccine, but at the same time various drugs (either via research or experimental use) that had been used for other viruses were tested, but it soon became clear that they were not an adequate response to the emergency. The vaccines, while failing to prevent the disease completely, significantly reduced its severity and prevented not only more deaths but also the collapse of the health system. Thus, at a time when there was a lack of certainty and knowledge, two goals could be identified:

1. effective vaccines that could end or minimise the pandemic,
2. an equitable distribution of vaccines globally.

Within a few months of the start of the pandemic, there were already good candidates as possible vaccines. Scientists quickly managed to find different types of proposals: sophisticated and new technology such as the mRNA vaccines from Pfizer or Moderna, or more traditional processes such as Oxford-AstraZeneca; other candidates such as Sputnik in Russia and different alternatives from China also appeared. Accelerated approval was also achieved in order to start production quickly. The first goal—developing effective vaccines—proved feasible.

However, the second goal—an equitable global distribution of vaccines—would prove a tricky, ongoing challenge. This objective extends beyond vaccine efficacy, reflecting power asymmetries and the possibility of access to valuable goods or resources.

Looking back on the contemporary history of vaccine distribution in global public health, the precedents are not promising. In 2009, the H1N1 virus killed almost 300,000 people. Within seven months a vaccine was developed. However, Australia, the US, Canada and six other high-income countries[2] accumulated most of the vaccine supply (Bollyky and Bown 2020). Mexico, which had been ravaged by this pandemic, had signed a pre-purchase agreement that was not honoured. It was only after the intervention of the WHO that 10% of what was withheld by these nine countries was released, in a tragically late response (Lie and Miller 2020).

Similar coordination failures and power imbalances emerged during COVID-19.[3] Countries or regions ravaged by the virus did not get the vaccines they desperately needed. Some nations purchased or reserved far more doses than were needed for their populations; donations were made to countries that did not have such urgency or could not distribute them, with the consequent expiry and destruction of the precious supplies (Emanuel and Persad 2021; Bollyky et al. 2021).

Empirical evidence illustrates this: from the beginning of the pandemic in 2020, we could observe the hoarding of supplies; for example, the US bought so much of the drug Remdesivir that the UK, which was in its first wave of COVID-19, had to wait three months to obtain this drug. By August 2020, the US had made seven bilateral agreements with six pharmaceutical companies for more than 800 million doses (Usher 2021). Europe had secured half a billion doses of vaccine through two deals and Britain had purchased 270 million doses (equivalent to 225% of its population) through five deals (Usher 2021). Canada, which made deals and purchased doses for almost nine times its population, had proportionally the most vaccines (Rastello and Bolongaro 2020). According to the *New York Times*, in March 2021, the US negotiated vaccines for 650 million people, almost twice its population. In addition, there were blockades of exports of vaccines and essential components (e.g., by Europe) (Lee 2021). In November 2020, a Duke University study pointed out that high-income countries bought 3.5 billion doses with an option for another 5 billion more and that most middle- and low-income countries would have to wait until 2024 to obtain vaccines. This same point was made repeatedly by the WHO (WHO, 2021).

---

[2] I follow the World Bank's denomination that divides countries into high-income, middle-income and low-income countries.

[3] I refer especially to the first two years since this chapter was written during this time, but this dysfunctionality and inequity remains.

A philosophical analysis leads us to liken this situation to a kind of Hobbesian state of nature in which countries find themselves in a struggle of 'every man for himself'. Disputes led to lack of coordination, competition between the different suppliers, and price increases fuelled by confidentiality clauses imposed by pharmaceutical companies. Added to this was the possibility of betting on products that did not work (for example, at the beginning (June–November 2020) it was not known which vaccines would be approved) or that the inputs would be delayed and inaccessible due to problems with production by pharmaceutical companies. This situation led to the over-purchasing of vaccines by high-income countries. But a handful of countries reaching vaccination coverage alone would not and did not solve social and economic problems. Strict border closures are impractical and economically damaging. Immunisation strategies fail when governments prioritise domestic needs exclusively. Not only does this modus operandi not prevent the proliferation of new strains, but the entire vaccination effort undertaken by a country may be in vain.[4] Moreover, these positions are not even economically sustainable. The ICC Research Foundation showed that up to $9.2 billion would be lost if access to vaccines was not guaranteed for low- and middle-income countries (ICC Research Foundation 2021).

3.3 A Prisoner's Dilemma?

From a game theory perspective, the attitude and actions of these high-income countries can be described as a prisoner's dilemma. This classic non-cooperative game (a bimatrix model) reveals tensions between individual and collective rationality. Its most classic formulation is by Howard Tucker.[5] According to this version, each person simultaneously chooses a strategy and the combination of strategies determines each person's gains. It posits the following scenario: two suspects are arrested and charged but lack sufficient evidence of guilt. If neither confesses they will be sentenced to one month in jail, if both confess they will be sentenced to 6 months in jail, but if one confesses and the other does not, the one who confesses will go free while the other will be sentenced to 9 months for the crime and for perverting the course of justice. The people involved in this game win overall if neither of them confesses. In theory we are in a non-cooperative game, so those playing cannot make commitments in a pre-game negotiation, nor can they agree. Thus, the only operative agreements would be those that are self-regulated (Sánchez Santos and Pena López 2003). And what it shows is that two people can be non-cooperative even if it is against the interest of both.

[4] For example, in early 2021 South Africa had purchased doses of the AstraZeneca vaccine that did not work in the country due to its own local variants. And something similar can happen to a country already vaccinated if a new strain is reintroduced that resists the immunisation given.

[5] The first formulation of the dilemma took place around 1950 and was the work of Merrill Flood and Melvin Dresher, later formalised by Tucker.

The situation between countries appears similar to a prisoner's dilemma, given that, to an important extent, global immunity is a public good[6] with which all countries should cooperate and from which all would gain. However, countries—like the individuals in the prisoner's dilemma—act out of self-interest to secure domestic immunity. This can lead to over-buying and not cooperating with global immunity. Thus, countries would never sacrifice immunising 25% of their population, or whatever is necessary to free themselves from the emergency,[7] and contribute the rest to a common pool. Or in this particular case, to allow other countries to buy or have access to vaccines as well. This, in turn, is counterproductive, given that, if there is no global immunity, in the long run, the country that hoarded the vaccine can suffer, as new strains can appear, trade will not function, there will be no tourism, and so on. In other words, pursuing self-interested behaviour has negative consequences.

However, empirical evidence shows that self-interested herd immunity fails to restore normality (even if borders were closed to countries without non-equivalent immunity situations). In July 2021 in different European countries that had relatively good vaccination rates, new strains did not allow a return to normality. Faced with the supposed vaccine passport and the opening for the northern hemisphere summer, for example, on 7 July 2021 Catalonia closed some activities and set limits on opening hours with the cost that this implied for the local economy and tourism. In relation to the above, it should be noted that, based on self-interest, the rational action of buying can justify nationalist positions. Yet cooperation is necessary for collective benefit.

However, from a theoretical perspective, the dilemma that countries were facing with the distribution of vaccines diverges from the standard model. What is presented in the traditional theory is a non-cooperative game: the players cannot make commitments in a pre-game negotiation, nor can they agree on anything (Sánchez Santos and Pena López 2003). However, one element to consider in this version of the prisoner's dilemma is that in the case of the different countries, it is those countries (as players of the game) who can modify the conditions to make cooperation possible. Recall that it is the high-income countries that are the main players in this game. If the countries with high impact and strong economic power (such as the US, the UK, Germany, Japan) had agreed upon and allowed an equitable and financed sharing (as they did successfully with the scientific development of vaccines),[8] the distribution of vaccines for which a mechanism already existed (COVAX), a possible solution could have been found and a way out of the present dilemma could have been achieved.

Unlike the standard dilemma, in the pandemic situation, it is the participants who can set the terms and conditions of cooperation and can even punish those who refuse to cooperate. Unlike classical game theory, real-world mistrust and betrayal could be mitigated. That is, if the issue of cooperation had really been on the international

[6] I am aware that there is debate as to whether immunisation should be considered an overarching public good, but it certainly shares some characteristics with public goods (Bernstein and Randall 2020; Barrett 2007).

[7] This position will be made explicit in the next section.

[8] Operation Warp Speed, aimed at promoting the rapid development of vaccines, was successfully implemented by the US.

agenda, if the logic of nationalism and self-interest had been set aside, and if we had thought cooperatively and transparently, the situation would have been very different and the dilemma could have been solved. While prior agreements were feasible, an entrenched non-cooperative tradition persisted.

3.4 Nationalism or Vaccine Cosmopolitanism?

Another issue that needs to be discussed in terms of ethics is nationalism in the purchase of vaccines. It was not just high-income countries that bought large vaccine stocks and implemented export bans. Middle- and low-income countries also desperately bought vaccines. However, they generally fared less well, not only because they were unable to get the coveted doses quickly due to their lower bargaining power, but also because of costs and prices. For example, South Africa purchased AstraZeneca doses at almost twice the cost of European countries (Moodley et al. 2021).

Is there any kind of ethical justification for these behaviours, and are there acceptable versions of them? Vaccine nationalists argue that every government has a moral duty to prioritise its own people before helping others abroad. They should promote and protect the rights and welfare of their own residents and citizens.[9] Normatively, governments have duties to their citizens. National governments have the capacity to mobilise and redirect resources and can be held accountable for failing to do so (Lie and Miller 2020, 8). In the absence of global governance, national governments have a primary responsibility towards their residents.

The most radical position is that of the extreme nationalists. For them, countries can negotiate multiple (and redundant) advance purchase agreements (APAs) and renounce their participation in a redistributive scheme. Their behaviour is driven by self-interest. This aligns with a statist view that prioritises national security and sovereignty (Wenham et al. 2022). Among the reasons for entering into bilateral negotiations is the need to accumulate large numbers of vaccines, given that global cooperative efforts in less than ideal circumstances are not guaranteed. There are no strong international mechanisms in place to target the distribution of health commodities. These are country decisions that are made on an individual and sovereign basis. Often these attitudes are also fuelled by domestic social pressure for vaccines.

However, with the COVID pandemic, the nationalist logic does not quite work. As noted, this pandemic would have ended more quickly if all countries had had access to a fair proportion of vaccines. The risk of new strains does not allow the extreme nationalist approach to succeed. In this respect, prudential reasons can be raised against it. Secondly, there are moral arguments: extreme nationalism deprives countries with fewer resources of access to vaccines in a reasonable time. It depletes

[9] There is an internal discussion regarding who is covered: only citizens or also residents? If the latter, a second approach is to discuss whether illegal residents, undocumented residents, etc. are also covered.

limited vaccine resources and in some cases wastes them. Thus, extreme nationalism fails the second goal outlined earlier—that of equitable global distribution of vaccines.

In contrast to these nationalist proposals, the opposite view is the cosmopolitan position: membership of a community, citizenship or associative ties are ethically irrelevant to questions of distributive justice. From political philosophy as well as from more traditional ethics, we can find those who defend these positions (Pogge 2005; Gargarella 2007; Luna 2007). An example of this type of approach is found in utilitarianism (Singer 1972). This position rejects reasons of relative agency or special duties for decision-makers in different countries. For example, Peter Singer points out that there is no doubt that we instinctively prefer to help those who are close to us. He specifies that few might stand and look at a drowning child, and yet many might ignore the preventable deaths of children in Africa or India. The question, however, is not what we usually do, but what we ought to do, and it is hard to find sensible moral justification in the view that distance or community membership makes a crucial difference to our obligations. (Singer 2011, 202–203).

What this kind of approach does is engage with the idea that all people are of equal value regardless of communal identities or belonging. Recall that in the utilitarian calculus each person counts as one and only one, and even my own interests do not outweigh those of a stranger (Mill 2017).

These arguments also apply to the global allocation of vaccines. The cosmopolitan position, which calls for collaborative work based on common health ideals, argues that allocation criteria should prioritise distribution to countries on the basis of need (e.g., considering the rate of deaths and infections) or in proportion to the size of a country's population. The cosmopolitan position may disagree on the principles and allocation scheme, but agrees that any allocation of vaccines should give equal consideration to people regardless of their national affiliations. It is also worth noting that it is the spirit of global proposals that nurtures international cooperation in terms of health. In this type of proposal, the values that are present are equity, solidarity and common goals to improve people's health (Wenham et al. 2022). In short, proponents advocate universal equity. And, in theory, this position would meet the second goal we were seeking to achieve. In this sense, it seems to respond ethically to the challenge posed.

It should be noted that in this approach, while the recipients are the people, the global distribution of vaccines is done via governments or nation states. So for the cosmopolitan model to work, it seems that we need a global authority or perhaps some kind of supranational mechanism that can impose terms and conditions on countries, as well as, perhaps, some limits on pharmaceutical companies. No supranational body held such authority during the pandemic. Thus, given the empirical and institutional obstacles and global dysfunctionality, it is very difficult to implement these proposals. In this sense, even if it is our Paradise Island, it is a position that is too difficult to put into practice, and it is still almost impossible to set foot upon it.

3.5 Varieties of Moderate Nationalism

In contrast to these positions, moderate nationalisms argue that countries have a responsibility to prioritise their own population, but with certain limitations. Prioritising one's own population is not incompatible with recognising certain duties towards all people. Moreover, it is wrong to ignore the suffering of other people when they can be helped with relatively little effort.

One point for debate is what a country is justified in doing. Which limits must be imposed? When the pandemic emerged, there were some defences of moderate nationalism but no concrete proposals. For example, Ferguson and Caplan argued for some nationalism, but explicitly stated that they did not claim to know what the limits of acceptable national priorities were (Ferguson and Caplan 2020). Miller and Lie (2020), who also advocated a moderate line of argument, pointed to COVAX as a middle way, but also did not establish criteria for how to balance national and international duties (Miller and Lie, 2020).

At least two versions of moderate nationalism could be reconstructed when it came to the issue of vaccines during the pandemic. The first is the herd immunisation approach—the attitude that countries have in fact implemented—whereby each country should vaccinate its population until what is known as herd or flock immunity is achieved (around 60–70% of its population, the exact number depending on empirical issues). This position supposedly allows the economy to reopen and pre-pandemic social life to resume, and it is suggested that countries can then donate surplus doses. However, achieving this herd immunity is not so easy—not only because of the large number of vaccines needed, but also because of the refusal of many people to be vaccinated. High levels of resistance to vaccination make herd immunity very difficult to achieve and the possibility of sharing vaccines very remote. In July 2021 France made vaccination compulsory for workers in health services or nursing homes, and a health pass mandatory for access to restaurants, cinemas, clubs and theatres, not only to promote adherence to vaccination, but also to obtain higher levels of immunity to protect against new strains such as the Delta variant.

The second type of moderate nationalism is the Fair Priority for Residents (FPR), more colloquially referred to as the 'flu risk' standard (Emanuel et al. 2021).[10] This approach, though more demanding, was a fairer way to counteract exacerbated nationalism and the overly lax approach of herd immunity, as it took seriously the idea that it is wrong to ignore the suffering of others and also set a lower cut-off line than the alternative approach. In essence, it implied a sacrifice on the part of the population as opposed to herd immunity. With this approach, governments can withhold sufficient doses to avoid a crisis level of mortality, ensuring that the health system can continue to function for people with other illnesses, and appropriate levels of intensive care can be maintained. It also allows for the resumption of some economic and social activities, but ensures that certain public health measures must be maintained. In other words, it does not imply a complete return to normality.

[10] So named because it takes as a reference the levels of deaths and dysfunctionality posed by an influenza epidemic (Emanuel et al. 2021).

For example, protective masks should continue to be worn and air filtration systems used in public buildings, schools and nursing homes. While it also involves empirical issues and modelling, it is estimated that it would involve vaccinating 25% and then continuing with 1% of the population per week.

The flu risk position argues that there is a moral obligation to other people who are suffering regardless of their national residence or affiliations, and that vaccines should be donated or made available to countries in need. It follows that this obligation arises at a relatively early stage of the country's own vaccination programme. It still implies certain sacrifices as it does not authorise a return to normality but this is a minor sacrifice compared to the deaths and suffering it can prevent for other countries. It aims to achieve the most equitable sharing possible, while taking into account national obligations in the crisis situation introduced by the pandemic.

3.6 Assessing the Different Proposals

We can now return to our initial philosophical analysis: how are we to get by in this non-ideal world, permeated by the constraints we have examined? How do we combine the ideal and the non-ideal? While it is true that these actions were taken in times of crisis and desperation and do not seem to meet the conditions of careful deliberation and planning, neither can they be taken as entirely arbitrary, as they do involve national considerations and reflections by governments. As such, they imply a certain view of what should be done and of the different ethical obligations of countries, and may to a greater or lesser degree appeal to ideal or non-ideal models.

Chapter 1 sets out a decision mechanism that requires some criteria to justify our choices between non-ideal theories (NITs) and ideal theories (ITs). This instrument has five requirements:

R1: The proposal should be politically possible;
R2: The proposal should be likely to be effective;
R3: The proposal should be morally permissible;
R4: Among the proposals to be compared, select the one where the most grievous injustices are dealt with or, where this is not possible, seriously considered.
R5: The proposal should appraise or consider long-term goals, in order to avoid endorsing strong obstacles to them when we try to move towards those goals.

Can we apply this decision-making mechanism appropriately in this case?[11]

If we consider the cosmopolitan position as the ideal theory (one that proposes to act cooperatively and in solidarity with the goal of distributing vaccines equitably among all countries), we see that the ideal theory, given the desperate situation and social pressure that existed during the first years of the pandemic, is unattainable. Not only were nationalistic tendencies exacerbated, but no international or multilateral body or mechanism existed that could decide on the global purchase and

[11] For a justification of each of the requirements see Chapter 1.

distribution of vaccines. The WHO is a global public health institution but has no binding powers: indeed, the WHO's *International Health Regulations* (IHR) were not seriously considered during the pandemic (Wenham et al. 2022).[12] In the next chapter we will look at some attempts or alternatives such as COVAX that tried a more idealised approach.

If we take the rest of the proposals as non-ideal theories, we can see that, with respect to the decision-making mechanism, radical nationalism would only fulfil the first requirement since it is politically possible (it was in fact the one that prevailed) but it does not respect any of the following criteria: not only is it not very effective but, above all, it does not fulfil either the third or the fourth requirement, as it is not morally permissible. It ends up in a Hobbesian state of nature with all countries fighting each other. Nor does it consider the most serious injustices such as the situation of countries being devastated by the pandemic. In this sense it seems clear that radical nationalisms do not pass the test of the five criteria. They must be rejected.

The key question is: which moderate nationalism is the most ethically defensible? Both the two remaining non-ideal positions appear to be politically possible (the first requirement). It has been shown that it is not easy for countries to be generous or wait patiently for precious vaccines, but if the world were to evolve towards acting less self-interestedly, both moderate proposals could be implemented. The herd immunisation position seems more politically feasible—it can cover more people, more palatable for politicians and citizens, and it can avert any danger of crisis—than the flu risk position. However, the second requirement—that the proposal be likely to be effective—does not seem to work with herd immunity. Given that the refusal of much of the population in wealthier countries to vaccinate makes it difficult to achieve the desired herd immunity, asking less, as the flu risk proposal does, may be more effective, easier to achieve and more acceptable to the population.

Further, if the goal is to minimise or end the pandemic, the flu risk option also seems more effective because restricting the use of vaccines earlier by countries with greater resources and sharing extra vaccines with countries with greater needs should restrict the overall pandemic and avert new strains more quickly. And if a country with resources and vaccines really needed them, it would be justified in using the vaccines in its population. The flu risk proposal appears to fulfil the third requirement of the decision-making mechanism, with a model in which more resource-rich countries are willing to restrict themselves in order to collaborate with the countries that are suffering the greatest need. The flu risk option again appears better with respect to the fourth requirement, as it places greater limits on countries with more vaccine resources and proposes earlier sharing, thus helping countries that are suffering from deaths or saturations in their intensive therapies more quickly. While both proposals seem better than the radical version, the flu risk model is more appropriate and of higher ethical quality.

Finally, we are left with the fifth requirement of our decision-making mechanism. This assesses how long-term goals are considered. Here it seems that moderate nationalism may be going in the right direction, but other instruments are needed

---

[12] This point will be returned to in Chapter 5.

when thinking about how to avoid obstacles that stand in the way of achieving the goal of ending the pandemic and the goal of equitable distribution of vaccines globally (the second goal presented at the beginning of the chapter).

3.7 Conclusion

This chapter presented the dysfunctional real world's impact on the equitable sharing of scarce resources such as vaccines. Non-cooperation was analysed in terms of the prisoner's dilemma, although it was also shown that, in some ways, such a dilemma should not have happened. Contrary to theoretical models, those who govern countries and make decisions could have agreed on a fair share.

The decision-making mechanism presented in Chapter 1 was applied and analyzed the ethical acceptability of the possible domestic approaches of each country with its nationalist and moderate visions. Finally, it was shown that although one of the options is better, the problem of fair distribution still requires further solutions.

References

Barrett, S. 2007. *Why cooperate? The incentive to supply global public goods*. Oxford University Press.

Bernstein, J., and P. Randall. 2020. Against the public good concept of public health. *Public Health Ethics* 13 (3): 225–233.

Bollyky, T., and C. Bown. 2020. The tragedy of vaccine nationalism. *Foreign Affairs* 99 (5): 96–108.

Bollyky, T. et al. (2021). Epidemiology not geopolitics, should guide COVID-19 vaccine donations. *The Lancet* 8 June 398 (10295): 97–99.

Emanuel, E., and Persad, G. (2021). This is the wrong way to distribute badly needed vaccines. *The New York Times*, 25 May, Section A: 21.

Emanuel, E., et al. 2021. On the ethics of vaccine nationalism: The case for the fair priority for residents framework. *Ethics & International Affairs* 35 (4): 543–562.

Ferguson, K., and A. Caplan. 2020. Love thy neighbour? Allocating vaccines in a world of competing obligations. *Journal of Medical Ethics* 47: e20.

Gargarella, R. 2007. La comunidad igualitaria y sus enemigos: Liberalism, republicanism e igualitarismo. *Perspectivas Bioéticas* 12 (23): 56–75.

ICC Research Foundation. (2021). *Study shows vaccine nationalism could cost rich countries US$4.5 trillion*. https://iccwbo.org/media-wall/news-speeches/study-shows-vaccine-nationalism-could-cost-rich-countries-us4-5-trillion/

Lee, G. (2021). Coronavirus: EU demands UK-made AstraZeneca vaccine doses. *BBC News*. https://www.bbc.com/news/world-europe-55822602.

Lie, R., and F. Miller. 2020. Allocating a COVID-19 vaccine: Balancing national and international responsibilities. *The Milbank Quarterly* 99 (2): 1–17.

Luna, F. 2007. Extrema pobreza y bioética: ¿Caridad, rescate u obligación? *Perspectivas Bioéticas* 12 (23): 13–31.

Mill, J. S. 2017. *Utilitarianism*. Hackett.

Moodley, K. et al. (2021). Hard choices: Ethical challenges in phase 1 of COVID 19 roll out in South Africa. *SAMJ* 26 March 111 (6): 554–558.

Pogge, T. 2005. World poverty and human rights. *Ethics and International Affairs* 19 (1): 1–7.
Rastello, S., and K. Bolongaro. 2020. *Canada has reserved more vaccine doses per person than anywhere*. Bloomberg. https://www.bloomberg.com/news/articles/2020-12-07/canada-has-reserved-more-vaccine-doses-per-person-than-anywhere.
Sánchez Santos, J., and J. Pena López. 2003. *La ética en la economía o el absurdo del dilema del prisionero*. Conference: Jornadas de Economía Crítica.
Schaefer, G., et al. 2023. The importance of getting the ethics right in a pandemic treaty. *The Lancet Infectious Diseases*. https://doi.org/10.1016/S1473-3099(23)00364-X.
Singer, P. 1972. Famine, affluence and morality. *Philosophy and Public Affairs* 1 (3): 229–243.
Singer, P. 2011. *Practical ethics*. Cambridge University Press.
UN. (2022). *Calling vaccine inequity 'biggest moral failure of our times', Secretary-General says focus must be on fairness, in COVID-19 tools accelerator launch remarks*, 9 February. https://press.un.org/en/2022/sgsm21137.doc.htm.
Usher, A. D. 2021. A beautiful idea: How COVAX has fallen short. *The Lancet* 397 (10292): 2322–2325.
Wenham, C., M. Eccleston-Turner, and M. Voss. 2022. The futility of the pandemic treaty: Caught between globalism and statism. *International Affairs* 98 (3): 837–852.
World Health Organization (WHO). (2021). WHO Director-General's opening remarks at 148th session of the Executive Board.*The World Health Organization*. https://www.who.int/director-general/speeches/detail/who-director-general-s-opening-remarks-at-148th-session-of-the-executive-board.

Chapter 4
Bilateral Donations and COVAX

Abstract The fourth chapter examines other pandemic-era mechanisms. It proposes alternatives that would have implemented the fifth criterion of the decision-making analysis to advance equitable global distribution of vaccines. It focusses on two initiatives that emerged during the pandemic. The first involved bilateral donations between individual countries, whereas the second was an international platform for researching, producing and distributing vaccines, known as COVAX. COVAX was an interesting and promising first step based on values such as solidarity and equity. However, it did not achieve an efficient and equitable distribution. This novel, complex mechanism is analysed. The chapter finishes by assessing both initiatives.

Keywords Ideal and non-ideal theories · Decision-making mechanism · COVID 19 pandemic · Vaccine global allocation · Bilateral donations · Fair priority model · COVAX

4.1 Introduction

As discussed in the previous chapter, the flu-risk standard (Emanuel et al. 2021) may satisfy four of the five criteria of the proposed decision-making mechanism, yet it remains insufficient. This chapter examines pandemic-era mechanisms and proposes alternatives that would have implemented the fifth criterion to advance equitable global distribution of vaccines. I assess two initiatives that emerged during the pandemic. The first involved donations between individual countries, while the second was a novel, complex approach: an international platform for researching, producing and distributing vaccines, known as COVAX.

F. Luna, *The Ideal and Non-Ideal in Bioethics*, The International Library of Bioethics 115, https://doi.org/10.1007/978-3-032-14594-9_4

4.2 Varieties of Moderate Nationalism and Bilateral Donations

A first option is to look at a phenomenon that emerged during the pandemic: bilateral donations. Moderate flu-risk nationalism could have been complemented by a commitment to bilateral donations. This sharing model could have resolved the crisis while respecting high-income countries' autonomy and improving vaccine access for resource-poor nations. But did something like this happen? Did this approach resolve the dysfunctionality and achieve the second goal outlined in the previous chapter, that is, equitable distribution of vaccines globally?

Before ethically assessing this sharing proposal, it is worth reviewing some of the so-called 'vaccine diplomacy' practised by some countries. The analysis discussed in this section applies only to individual arrangements between countries. One issue to bear in mind is that those countries that can engage in this type of behaviour are generally those that produce and have easier access to vaccines. But it could also be those that have purchased far more vaccines than they need (e.g., Canada).

Bilateral donations began to emerge in late 2021. China became a major donor, sending 23 million doses to 72 countries. According to Kiernan et al. of Think Global Health, only two of these 72 countries did not belong to China's Belt and Road Initiative (Kiernan et al. 2021). These researchers point out that in making the donations, China was seeking to expand and deepen cooperation by those countries with its One China policies against Taiwan, Hong Kong, Tibet etc.

India also made donations and had a similar policy. It distributed 10.7 million doses to 48 countries. However, three-quarters of the donations, i.e., 8.2 million, were distributed to only 11 countries in Southeast Asia, with other countries only receiving 70,000 doses, and in general the criterion was to deliver vaccines to the same countries as China in order to counterbalance China's leadership (Kiernan et al. 2021). Regions such as Latin America or Eastern Europe, which were in a critical situation in the first half of 2021, received virtually no donations. Southeast Asia accounted for 55.59% of donations but had only 20.73% of confirmed COVID cases.

Thus the structure of bilateral donations deepens power relations and maintains a logic of influence and geopolitical interests. In this sense, not only does it foster arbitrariness and self-interest in some countries, it also proves highly inefficient. Inefficiency worsens when countries donate surplus or expiring vaccines, as it takes time to effectively distribute these supplies in each country (the operation is complex and requires some time to organise). This type of bilateral donation mechanism does not seem to meet the second goal of equitable global distribution. It does not consider those who really need to vaccinate their populations due to the incidence or crisis they are suffering, and in this sense it is problematic.

A possible justification based on principles of global justice could cite historical ties or friendships.[1] It could perhaps be seen as an aspect of rectificatory justice.

[1] I am grateful to Romina Reckers for pointing out this possible counter-argument.

However, this does not seem to be the right strategy for the kind of problem faced. If one wants to end the pandemic, one should first relieve the countries with the highest number of cases or in the worst situation (not only on humanitarian grounds, but also because such a situation can generate strains that can spread to several countries). Thus, the right thing to do would be to favour the countries most affected by the virus, even if this means setting aside historical or friendly relations. Moreover, returning to our decision mechanism, it does not meet the second requirement (i.e., that it is likely to be effective), nor does it seem to meet the fourth requirement (that it requests consideration of the option in which the most grievous injustices are addressed or seriously considered: countries badly hit by the virus were ignored).

While countries could apply equity criteria and avoid donations by being self-interested, the temptation to obtain other benefits from donations is very strong. There is no obligation or benchmark to follow in such practices—it remains an individual decision for each country. On the other hand, from an ethical perspective, this modality appeals to charitable models that rightly depend on good intentions (if any) and self-interest. The fifth requirement of the proposed decision-making mechanism is also not fulfilled, as long-term objectives must be considered or evaluated in order to avoid creating strong obstacles to them when trying to move towards them. For all these reasons, it does not seem to be the best model to follow.

4.3 COVAX: A Promising New Approach?

Another type of proposal was the COVAX mechanism.[2] Towards the beginning of the COVID-19 pandemic, in April 2020, the WHO, the European Commission, and some international leaders proposed a multilateral scheme: the ACT (Access to Covid-19 Tools) Accelerator, this mechanism was closed in December 2023. One of its three pillars was the Global Access Fund for Covid-19 Vaccines (COVAX). It was led by GAVI (Global Vaccine Alliance), CEPI (Coalition for Epidemic Preparedness Innovation), and WHO, together with other stakeholders. COVAX, which involved 190 countries, had three main objectives. The first was vaccine research. COVAX funded several vaccine platforms, mostly during 2020 but continuing in 2021 (when some vaccines such as Pfizer, Moderna, or Oxford-AstraZeneca had already been developed), looking for new alternatives that were better than existing vaccines—such as single-dose vaccines, vaccines that did not require a cold chain, or other modes of delivery, for example, oral vaccines. Secondly, equitable global distribution was proposed. Finally, the platform included the expansion of manufacturing capacity, technology transfer and vaccine production.

[2] From October 2020 until 31 March 2021, I was the principal investigator of a WHO-funded project to evaluate the first phase of COVAX. In addition to the final report, we published several articles with the project team, and several of the ideas I propose here are drawn from that literature. Luna and Holzer et al. (2021); Manríquez Roa et al. (2021); Holzer et al. (2021); Holzer et al. (2022).

Types of participation differed among countries. Self-funded countries, including high-income but also many middle-income countries, made an upfront payment. Most Latin American countries fell into this category (Argentina, Peru, Paraguay, Guatemala, etc.). A second category, the Advanced Market Commitment (AMC), included 92 low- and middle-income countries that received donated vaccines. Note that only four countries in Latin America were considered: Bolivia, El Salvador, Honduras and Nicaragua. This point was important because most Latin American countries should have been included in the range of countries that were expected to be self-financing.[3] This was notable because it showed that many other low- and middle-income countries in the region—such as Paraguay, Ecuador and Guatemala—were overlooked.

One of COVAX's main objectives was to procure vaccines for 20% of each country's population.[4] The proposal was to allocate at least 2 billion doses by the end of 2021. The first delivery of vaccines to Ghana took place on 23 February 2021.

How can we assess this mechanism, as both an organisation and a project? As a project, it was the first time in history that a mechanism for equitable allocation of vaccines had been established at the global level. Until then, there had only been partial models, such as the Pan American Health Organization (PAHO) Revolving Fund which, though successful, had only involved regional distribution. Here it is important to highlight the significance of breaking with historical thinking in moving away from Hobbesian logic and attempting to implement an equitable global approach. This was a significant step forward. In representing progress and a break from the status quo, this should give us hope, although I will argue later that there remains room for improvement. Another element to keep in mind is that all countries were able to participate, not just those in need of funding, as COVAX aimed to cover most countries. Furthermore, at the beginning of the pandemic, when vaccines were still in the research stage, COVAX presented a compelling proposal for access. This was because, had vaccines been procured through COVAX, purchases would have been made generically from a common pool, and then those that were approved would have been distributed—thereby avoiding the risk of investing in vaccines that might have failed, or the need to buy from multiple pharmaceutical companies in case one did not obtain the necessary approvals. In this sense, the proposal was highly rational and efficient, as it aimed to avoid superfluous purchases that could lead to stock-outs. It should also be noted that it was implemented quite quickly (in less than a year it was distributing vaccines). From an ethical perspective, the mechanism prioritised the poorest countries through AMC donations, and WHO played an important role in the efficient planning of immunisation strategies. Lastly, the mechanism allowed for supplemental bilateral agreements, which should have allowed for more equitable vaccine distribution and put less pressure on the mechanism. This last feature, which could be considered positive, turned out to be a double-edged sword. While it reduced the pressure on COVAX, as it was only one of the available means of

---

[3] This is in response to the World Bank's classification, which is not without its problems and criticisms.

[4] But it also allows self-financing countries to buy up to 50% for their population (Usher 2021).

accessing vaccines, it helped reduce the consequent risk of manipulation, corruption or bias from more powerful countries in their efforts to secure more vaccines, unfortunately it also enabled extreme nationalism to flourish by failing to curb the excesses and ambitions of certain countries.[5] The fact that some countries purchased and stockpiled excessive quantities of vaccines proved completely counterproductive for COVAX, which did not manage to obtain vaccines in time.

There were many positive elements in the establishment of a mechanism such as COVAX that is both necessary and inspiring. If, however, one looks at its concrete implementation, one can ask whether it really fulfilled the second goal proposed in the previous chapter, "equitable distribution of vaccines globally". Not everything turned out as expected, and it was tarnished by criticisms of various kinds. If a mechanism such as this is to achieve an equitable distribution of vaccines or medicines, it is in need of serious tweaking.

4.4 The Proportional Allocation Criterion

If, instead of evaluating COVAX as a project, one looks at its concrete implementation, one can ask whether it really fulfilled the second goal proposed in the previous chapter: "equitable distribution of vaccines globally." After a positive first glance, we see that this promising mechanism was not without its drawbacks. A central issue, present in the COVAX design from the outset, was the criteria for distribution and how equitable it would prove to be beyond goodwill statements. COVAX proposed what was known as a Proportional Allocation Standard (PAS). This meant that it would distribute an amount proportional to the population of each participating country, up to 20%. Only once this initial 20% had been reached would the risks or vulnerabilities of each country be taken into account (Emanuel et al. 2020). COVAX outlined three stages: starting with 3% (mainly to cover health workers), a second stage up to 20%, and finally 60% in the third stage, considered necessary to reach herd immunity with the vaccines. At first glance, this proportional system of distribution seemed fair. It offered a formal answer to the problem of justice and appeared to follow an egalitarian moral approach. It had the advantage of proposing a relatively simple mode of distribution that admitted neither arbitrariness nor bias. It was supposed to distribute the doses necessary to cover the estimated percentages of people in the proposed stages. Indeed, one might say it provided a clear and concrete baseline standard.

But a more thorough analysis revealed problems with this model. The pandemic did not affect all countries equally and, as in other health circumstances, it seemed fair to respond to similar needs in a similar way. Moreover, as was later seen, early access

[5] Note that as late as February 2022, most countries had already passed the major waves of severe infection. PAHO noted that "Although 63% of the population in Latin America and the Caribbean has already been vaccinated for COVID-19, coverage remains uneven, with 14 countries and territories that have not immunised 40% of their population. Haiti remains at the bottom of the list, with vaccination coverage of less than 1.2%." (PAHO 2022).

to vaccines was vital to resolving the crisis. Thus, it could be argued that it would not make sense to allocate antiretrovirals for HIV-AIDS according to population size, but rather according to the burden of the disease in each country. For example, the situation in sub-Saharan Africa was very different from that in Western Europe. Nor would it be an appropriate criterion to distribute wheelchairs or chemotherapy treatments where they were not needed. Similarly, allocating vaccines proportionally did not seem to be the best solution. A concrete example illustrates this point more clearly. At the end of September 2020, Peru, with its population of approximately 32 million, had more than 800,000 cases of COVID-19 and 32,000 deaths, while Poland (with a population of 38 million) had only 89,000 cases and 2500 deaths. The COVAX proportional system would have distributed more doses to Poland than to Peru—even though Peru had nine times more cases and thirteen times more deaths.

In addition to the formal approach to ethical and efficient vaccine distribution, procedural issues should also be taken into account. For example, the speed with which a vaccine was distributed proved crucial. This was not adequately considered and may have implied biases or arbitrariness. Thus for Argentina, access to vaccines at the beginning of 2021—in February or March, before winter began—was not the same as it was in June. In March 2021, Peru continued to face a difficult situation: by that point, it already had 1,400,000 cases of COVID-19 and 48,500 deaths, while Ghana (with a similar population) had only 86,000 cases and 700 deaths. Even accounting for potential under-reporting, the difference between the two countries was substantial. Yet the COVAX proportional system distributed first to Ghana rather than to Peru, even though Peru had sixteen times as many cases and nearly 70 times as many deaths.[6]

While one can sympathise with the priority given to Ghana and Africa in general, arguing from a logic of rectificatory justice based on the long and traditional neglect of these populations, at the time, it was not the country that needed the most vaccines in the face of the health collapse, deaths and increased poverty that an acute wave of cases implied, consider also vaccines' shortage and the desperation of many countries. Priority should be given to the countries that are most in need, and the proportional model does not prevent avoidable deaths. If COVAX's objective was ending or minimising the pandemic as far as possible, it did not take the right steps. And it failed on the criterion of fairness.

Delivery time is also highly relevant. In May 2021 Sudan destroyed 60,000 doses donated by COVAX and Malawi another 20,000 doses because they could not distribute them in time and the vaccines were expiring (Emanuel and Persad 2021). Côte d'Ivoire was able to distribute less than a quarter of the 500,000 doses it received from COVAX. But the problem of implementation is not unique to low-resource countries. Canada ran the risk that 600,000 AstraZeneca vaccines that COVAX sent there would also expire, given the debate about the safety of this type of vaccine that the country raised. Here it is not only the distribution model that fails but also the

[6] Although Ghana obtained vaccines through the donation system and Peru did not, this distinction should not be relevant for efficiency purposes.

back-up and screening that the WHO should do before allocating and shipping the vaccines.

Meanwhile, Latin America's situation was deteriorating. By 15 June 2021 Brazil had reached 500,000 deaths with more than 70,000 cases per day. Vulnerability and precariousness in the region led to excess mortality (Rekers and Luna 2023; Karlinsky and Kobak 2021; Lima et al. 2021; Yabar 2023). Specialised personnel and beds were also in short supply (Benza and Kessler 2022: 34).

4.5 The Equitable Distribution Criterion

A second proposal, the Fair Priority Model (FPM), emerged in early 2020 when there was still no ethical global distribution model and it was not known what COVAX's position would be. Before the pandemic, most concrete proposals for just distribution were implemented at a domestic level. At the initiative of Ezequiel Emanuel, a group of 18 academics from different countries met for several months to discuss the issue, publishing their findings in *Science* in early September 2020 (Emanuel et al. 2020).[7]

At the core of the FPM was the vaccine need criterion. This approach was based on three fundamental values. The first was benefiting people and limiting harm, a value widely recognised by different ethical theories. The main idea was to consider people and their interests, which required the relevant benefits to be defined and measured, and their relative urgency established.[8] Second, prioritising the disadvantaged. This is also a fundamental value in ethics and global health and requires the establishment of different types of disadvantage.[9] And a third, equal moral concern, i.e., not discriminating based on morally irrelevant differences (colour, sexual identity, etc.).

This model is structured in three stages. The aim of the first phase is to reduce premature deaths, other direct health impacts (e.g., permanent organ damage associated with COVID-19), and indirect health impacts (e.g., those caused by a collapsed health system or by people fearing to be checked or treated for other health problems). The idea is to prioritise those countries most in need of the vaccine. This stage ends when the health emergency is over and countries are faced with a health challenge similar to the possibility of a flu outbreak.[10] The second stage is to reduce the health problems that remain but also to take into account the serious economic and social deprivation caused by the pandemic. For example, it considers how having the vaccine can reduce the poverty generated by the pandemic by allowing resumption

[7] I was a member of the group who wrote that article.

[8] A successful vaccine would produce direct benefits by protecting people from death and morbidity from COVID-19; provide indirect benefits by reducing morbidity and mortality from a collapsed health system; and reduce poverty and socio-economic challenges by enabling economic, social and cultural re-opening.

[9] This is not a minor point: the article asks whether the countries that are worst off are those with the greatest poverty or, for example, those whose inhabitants have the worst life expectancy (Emanuel et al. 2020).

[10] I will return to this point in the next sections.

of activity, regenerating non-essential jobs, reopening schools, etc. The goal of this second phase is to return each country's poverty gap to pre-pandemic status. And the third and final phase is to reduce community transmission and return to pre-pandemic freedoms (Emanuel 2020).

Equitable distribution of the COVID vaccine is an issue of distributive justice, and although states or governments may be the initial recipients of the vaccines, the moral concern of this proposal lies with the ultimate recipient, i.e. the individual person. Thus, as well as prioritising those countries suffering most during the pandemic, the primary consideration was people's need rather than deservedness.

Another difficult point to resolve was the question of which metrics to use. For example, measuring the health infrastructure of each country would have been extremely relevant, but there is no specific global index that is complete and up to date. The WHO has an index of intensive care beds, but it is incomplete, and many countries do not record this data. In order to solve this type of problem, several metrics were proposed, taking into account the following factors: (i) faithfulness to established ethical values; (ii) simplicity; (iii) previous uses in global health and development; (iv) the possibility of obtaining estimates quickly as the pandemic evolves; and (v) sensitivity to harms that are difficult to measure directly. Instead of considering the number of deaths of individuals, SEYLL (standard expected years of life lost) was considered, which calculates the years of life that would be avoided by vaccine doses compared to a fixed reference standard that would be the same for everyone. One of the advantages of this measure is that, while all deaths are important, early deaths are particularly relevant (these, moreover, are the deaths that occur most in developing countries as they are the ones with the youngest populations). It thus functions as a proxy for prioritising the most disadvantaged and treats all people equally because it is a global metric that does not take into account differences in life expectancy between countries. For the reduction of serious socio-economic deprivation, in addition to the SEYLL that would be avoided, the reduction of absolute poverty, measured by the poverty gap and the incidence in gross domestic product, was incorporated, and for the last stage the transmission rates of the different countries were taken into account (Emanuel 2020).

The latter proposal, even if not followed by COVAX,[11] has many points of contact with the proposal made by the WHO Strategic Advisory Group of Experts on Immunization (WHO 2020).[12] Although our model is more complex than the Proportional Allocation Standard, it attempts to adequately meet the second goal identified: equitable distribution of vaccines globally.

4.6 COVAX: Pharmaceutical Companies and the Lack of Regional Development

COVAX could not be efficient even by its own standards. The balance sheet for 2021 (when global infections peaked) shows that it managed to deliver only a third of what was promised (approximately 910 million of the two billion doses originally proposed). In December, 310 million doses were distributed, with the practical problems of allocation, concrete application and overload for the health system that such a large quantity implies. Time is needed for effective implementation to reach people's arms (Taylor 2022).[13]

According to Borges and his colleagues, "By the end of 2021, ten major pharmaceutical companies were expected to deliver more than 12 billion vaccine doses. If distributed equitably, this volume of vaccine doses would have been sufficient to approach a global herd immunity." (Borges 2022: 3104). And they point out that extreme nationalism was one of the factors in the unequal distribution to COVAX, as several high-income countries such as the US, Canada, the European Union and Israel signed APAs to guarantee their citizens' immunity with repeated doses of the vaccines, even before the vaccines were approved for sale. By the time they could be marketed, these countries had agreed 38 APAs (US and UK 9 each, Canada 8, EU

---

[11] COVAX proposed its distribution model in September 2020, a few days before our first paper on FPM was published in *Science* (Emanuel et al. 2020).

[12] WHO identifies six principles that should guide global distribution between countries and at the national level. These principles are (1) human well-being, the first of which aims to reduce deaths and the burden of disease, the second to reduce social and economic disruption, and the third to protect the functioning of essential services; (2) equal respect, that is, treating all interests of individuals and groups as having equal moral consideration; and (3) global equity, which aims to ensure that the distribution of vaccines takes into account the social epidemic, risks and needs of all the countries to ensure that vaccine distribution takes into account the risks and needs of countries experiencing the greatest burden and to ensure that all countries commit to meeting the needs of countries that are unable to obtain vaccines on their own (both objectives mention low- and middle-income countries). This value framework is more in line with our proposal than that of COVAX. Its principle of human welfare is similar to our value of 'benefiting people and limiting harm'. The principle of equal respect matches our approach of equal moral concern. And finally the principle of global equity could also be read as taking into account the most disadvantaged. But what is very relevant is that in their explanation of the principles, they consider the same elements that we also point out as relevant: burden of disease, deaths, reducing social and economic disruption as well as taking into account risks, needs and vulnerabilities, among others (WHO 2020).

[13] A smaller but predictable quantity that can be sustained over time is often preferable to a large quantity that is sent all at once and can cause delays and distribution difficulties.

7 and Israel 5). This gave them priority access. The second factor Borges points out is the role of the pharmaceutical companies, as several of those that had agreed to distribute vaccines to COVAX prioritised selling them to countries that paid higher prices, delivering the coveted doses to COVAX towards the end of the mid-term or directly towards the end of the pandemic (Borges 2022: 3108). For example, Moderna charged COVAX US$7, the US government US$15 and the European Union US$18–35 (i.e., the average price was US$32 for high-income countries). Janssen, on the other hand, did not propose such a difference (it charged COVAX US$7.5 and the US US$10) (Borges 2022). For a more complete picture, it should also be noted that some of these countries had funded the research. For example, the US and Germany contributed US$2 billion and US$1.5 billion respectively, reaching almost US$5.9 billion by March 2021 (Borges 2022: 3102). COVAX could have established a transparent mechanism to recognise a certain priority for those countries funding the research that had helped discover the desired vaccines and therefore contributed to the common good. It could have established clear guidelines: giving priority to funding countries until they could emerge from the crisis or even achieve herd immunity. This would have given up to about 60% to these countries and not 900% or 700% as in fact happened. Such a solution would have equitably rewarded development assistance, rather than the arbitrary, self-interested and irrational bargaining that prevailed and which, in the end, boycotted COVAX by not achieving the desired doses. The result was an ineffective mechanism. As Borges and colleagues clearly show, both extreme nationalism and the self-interested, maximalist behaviour of pharmaceutical companies had significant consequences in the failure of equitable global sharing.

Contributing to the problems were shortage of funds to commit to purchase of vaccines, lack of coordination and fierce nationalistic competition, but also lack of regional capacity building and the bet on India and the Serum Institute as the main supplier of the Oxford-AstraZeneca vaccine. In April 2021 the Indian government blocked exports following the terrible wave of cases due to the Delta variant. India did not allow exports again until November, which also explains much of the shortage problems (Taylor 2022). You cannot have just one supplier.

A fundamental point related to the above which, in my view, was not adequately considered and which should be strengthened and expanded was the production of vaccines. This was to be done by CEPI in COVAX. With a global vaccine shortage and difficulty producing vaccines, one of the objectives of COVAX was to expand vaccine manufacture (Holzer et al. 2022). While it was undoubtedly valuable to move away from the traditional well-resourced industrialised countries to expand production and consider a country like India, it would have been better to spread production across several countries with some development experience.[14] COVAX should have diversified the development and production of vaccines by supporting not only India's Serum Institute, but also other regional centres with some technological expertise.

[14] It is important to note that countries that can produce vaccines must have some scientific background.

One of the objections to this idea was that vaccine production training could not be implemented quickly. Nevertheless, if work with several centres around the world had been started in 2020, this would have been achieved; and if one centre was unable to supply for some reason (as was the case when COVID-19 struck India), another of the already trained centres would have been able to take over and fulfil the orders. Producing new vaccines such as mRNA vaccines involved a series of intermediate steps that needed validation and a new learning process. But some countries already had experience with more traditional vaccine methods, and at least a start could have been made with these while training on more complex processes was under way. In fact, in Latin America, Argentina, Brazil and Mexico were all involved in vaccine production. Argentina supplied the active ingredient for AstraZeneca and Mexico the filling and termination process (Loewy 2020); and Brazil was taken as a pilot for mRNA vaccines by PAHO (Pan American Health Organization (PAHO) 2021). If these countries had been trained and supported from the beginning, perhaps the development of the pandemic in the region would have been different. Finally, it should be noted that there are strong interests operating behind the scenes, not only from pharmaceutical companies but also from some central countries supporting their industry.

Beyond these issues, we must consider the ethics of genuine regional capacity building, involving not only diversifying and avoiding dependence on industrialised vaccine-producing countries (generally the US and some European countries) but also promoting regional development and self-sufficiency. It entails moving away from the charity model to a model of greater autonomy that gives more sovereignty to middle-income countries. It was, after all, middle-income countries that were most left behind in this whole mechanism and international distribution (Emanuel and Jung, forthcoming). While wealthier countries bought and stockpiled vaccines and poorer countries received donations,[15] middle-income countries were left at a severe disadvantage. South Africa, for example, complained that it was charged twice the price that was paid by Europe for the AstraZeneca vaccine (Moodley et al. 2021). If technological development had been actively encouraged, there would have been greater respect and consideration for, and empowerment of, these intermediate and regional economies. However, it must also be recognised that this would not have been a panacea, as only a few countries are at a stage of technological development that would have allowed for further capacity building.

There were other problems in COVAX that need to be improved on. An institution like COVAX needs to be more transparent and have a better communication policy. There was also a geopolitical view that regards the Western world, led by the US, Europe and other nations as privileged (e.g., by prioritising their vaccines and industries). Solving these problems would provide confidence in the mechanism and would also gain more support, not only at the level of countries, but also from civil society. Most of them—transparency issues, training issues and the distribution model—can be solved in the short term.

---

[15] However, there were very few. As noted, in Latin America, COVAX only took Bolivia, El Salvador, Honduras and Nicaragua as countries to receive donations.

Thus, COVAX had several important shortcomings, but if implemented with necessary changes, would remain promising.

4.7 Re-Evaluating Options: COVAX Plus

So far we have presented what happened in practice during the first two years of the pandemic. And in the face of extreme nationalism, it was argued that the least harmful position was moderate flu-risk nationalism. However, such a proposal only stops or tempers the unbridled ambitions to hoard vaccines by countries that can do so.[16] But it is a minimal approach that does not go far enough; for example, it does not provide a mechanism to distribute vaccines that will not be used. The fifth requirement of the decision-making mechanism, which assesses long-term objectives as well as the avoidance of possible obstacles to such objectives, could not be met.

Both the bilateral donor model and the COVAX mechanism could complement the moderate nationalism approach, as they were developed in this non-ideal world. It has also been seen that both proposals pose difficulties.

Moreover, as we have seen, the bilateral donation model between countries is problematic. It does not meet the second requirement because it is not effective: it donates to those it wants to influence and does not take into account the needs of the different countries. Faced with an extremely scarce resource such as vaccines, this type of policy implies a highly questionable loss of efficiency. With regard to the moral permissibility of the third requirement, it is the self-interest of the benefactor country that lies behind the donations. To analyse the fourth requirement, we must compare this proposal with its alternative: the COVAX mechanism.

COVAX, as we have just seen, suffered from a number of problems: it failed to deliver what it promised and it did not adhere to the best model of justice (the FPM). However, it should also be remembered that it started from scratch (with no money or staff); despite its imperfections, it distributed the vaccines it obtained to a wide variety of countries and donated vaccines to those with fewer resources. Its proportional model was intended to avoid bias and in some ways to be more transparent. Its proposal and ambition was to be able to set itself up as the global distributor of vaccines, which seems to comply with the long-term goal of achieving a fair mechanism of universal distribution. For these reasons, I believe that COVAX, even with its shortcomings and problems, is better placed than a mere bilateral donation.

Are these outcomes acceptable? Our decision-making mechanism also allows us to consider the ideal. And the ideal is the cosmopolitan model in which distribution matches the need of each country. A fundamental improvement in the transparency of the COVAX mechanism would involve distribution according to the Fair Priority Model (FPM). Adequate funding, which seems to have been lacking at the

---

[16] While the industrialised countries with the most resources were the ones that reserved and purchased the most vaccines, all those that could afford to do so were also in the race.

height of the crisis (Holzer et al. 2022), would also be important. These modifications seem relatively affordable and do not require structural changes.. A COVAX that adopts these changes—which we could rename COVAX Plus—implies a superior COVAX that is closer to that ideal world that distributes much-needed medical supplies equitably among nations while also promoting regional development and independence.

Can similar reasoning and feasible improvements be applied to the bilateral donor model? It seems extremely difficult. Changing the political will of China, India or the US seems neither factually nor politically possible. Moreover, the charity model per se is arbitrary: you benefit whom you want, how you want. In the light of the above, it should be pointed out that COVAX is ethically superior, but that it must be improved and that it is COVAX Plus, or a similar mechanism, that should be implemented.

4.8 Conclusion

In this chapter, we have looked at some of the alternatives that were, in fact, available to achieve the long-term goal presented: an equitable global distribution of vaccines. While COVAX was shown to be ethically superior to bilateral donations, it had significant shortcomings and problems. I therefore propose COVAX Plus, or a similar mechanism, as an ethically superior alternative, which should incorporate a different delivery model, be more transparent, and effectively develop training and vaccine production at the regional level.

Though COVAX Plus offers an improvement, there is still a long way to go. The vaccine distribution left a lot to be desired, in particular regarding the questionable behaviours of the big players: rich, powerful countries and the pharmaceutical industry that were not up to the moral task. In the next chapter, we will address proposals far removed from what actually happened, but which emphasise the need to change the way the world functions and offer effective and equitable solutions.

References

Benza, G., and G. Kessler. 2022. The impact of the pandemic on Latin America: Social setbacks and rising inequalities. In *Persistence and emergencies of inequalities in Latin America: A multidimensional approach*, 33–49. Cham: Springer International Publishing.

Borges, L. C., H. Z. de Menezes, and E. Crosbie. 2022. More pain, more gain! The delivery of COVID-19 vaccines and the pharmaceutical industry's role in widening the access gap. *International Journal Health Policy Management (IJHPM)* 11 (12): 3101–3113.

Emanuel, E., and G. Persad. (2021). This is the wrong way to distribute badly needed vaccines. *New York Times*, May 24.

Emanuel, E., et al. 2021. On the ethics of vaccine nationalism: The case for the fair priority for residents framework. *Ethics & International Affairs* 35 (4): 543–562.

Emanuel E. J., and M. Jung (in press). How unfair and inequitable has the allocation of COVID vaccines been?

Emanuel, E. J., et al. 2020. An ethical framework for global vaccine allocation. *Science* 369 (6509): 1309–1312.
Holzer, F., et al. 2022. Charity or empowerment? The role of COVAX for low and middle income countries. *Developing World Bioethics* 23 (1): 1–8.
Holzer, F., et al. 2021. A matter of priority: Equitable access to COVID-19 vaccines. *Swiss Medical Weekly* 151: w20488.
Karlinsky, A., and D. Kobak. 2021. Tracking excess mortality across countries during the COVID-10 pandemic with the world mortality dataset. *eLife* 10: e69336.
Kiernan, S. et al. (2021). The policy of vaccine donation and diplomacy. *Think Global Health*, June 4, en: https://www.thinkglobalhealth.org/article/politics-vaccine-donation-and-diplomacy
Lima, E. E. C., et al. 2021. Investigating regional excess mortality during 2020 COVID-19 pandemic in selected Latin American countries. *Genus* 77 (1): 30.
Loewy, M. A. (2020). Argentina & México se unen para fabricar una vacuna "a riesgo" contra COVID-19. *Medscape*. https://espanol.medscape.com/verarticulo/5905794#:~:text=Compa%C3%B1%C3%ADas%20farmac%C3%A9uticas%20de%20Argentina%20y,con%20la%20financiaci%C3%B3n%20del%20proyecto,con%20la%20financiaci%C3%B3n%20del%20proyecto
Luna, F. 2007. Extrema pobreza & bioética: ¿Caridad, rescate u obligación? *Perspectivas Bioéticas* 12 (23): 13–31.
Luna, F., and F. Holzer. 2021. International cooperation in a non-ideal world: The example of COVAX. *Cadernos Ibero-Americanos de Direito Sanitário* 10 (3): 199–210.
Manríquez Roa, T., et al. 2021. Expert views on COVAX and equitable global access to COVID-19 vaccines. *International Journal of Public Health* 66: 1–8.
Moodley, K., et al. 2021. Hard choices: Ethical challenges in phase 1 of COVID 19 roll out in South Africa. *South African Medical Journal (SAMJ)* 111 (6): 554–558.
Organización Panamericana de la Salud (OPS). (2022). *La OPS entrega cien millones de vacunas de COVAX contra la COVID-19 en América Latina & el Caribe*. www.paho.org/es/noticias/15-2-2022-ops-entrega-cien-millones-vacunas-contra-covid-19-america-latina-caribe
Pan American Health Organization (PAHO). (2021). *PAHO selects centers in Argentina, Brazil to develop COVID-19 mRNA vaccines*. https://www.paho.org/en/news/21-9-2021-paho-selects-centers-argentina-brazil-develop-covid-19-mrna-vaccines
Pan American Health Organization (PAHO) (2022). *PAHO delivers 100 million COVAX vaccine doses to Latin America and the Caribbean*. https://www.paho.org/en/news/15-2-2022-paho-delivers-100-million-covax-vaccine-doses-latin-america-and-caribbean
Rekers, R., and F. Luna. 2023. Pandemic justice for and from Latin America. *Ethica* 22 (1): 1–25. https://doi.org/10.5007/1677-2954.2023.e97435.
Taylor, A. 2022. Covax vaccine deliveries surge in final stretch of 2021, with a record 300 million does sent out in December. *The Washington Post*. January 1, 2022. www.washingtonpost.com/world/2022/01/01/covid-covax-doses-delivered/
Usher, A. D. 2021. A beautiful idea: How COVAX has fallen short. *The Lancet* 397 (10292): 2322–2325.
WHO. 2020. *WHO SAGE values framework for the allocation and prioritization of COVID- 19 vaccine*. 14 de septiembre, 2020. Ginebra.
Yabar, C. 2024. Extreme poverty first: An argument on the equitable distribution of the COVID19 vaccine in Peru. *Developing World Bioethics* 24 (2): 97–101.

Chapter 5
Dynamic Obligations: Patents and International Treaties

Abstract This chapter proposes the possibility of implementing dynamic obligations and brings in a hint of the ideal. Dynamic obligations are aimed at transforming society. This chapter deals with a proposal that could not be introduced: a patent waiver policy. Although an attempt was made to implement it, the response came too late. This chapter examines its implications. Continuing with other possible dynamic obligations, the chapter presents the draft versions of the international agreement signed by the World Health Organization (WHO). The many theoretical, practical and political challenges facing this new instrument are reviewed. Dynamic obligations show how the ideal must inform assessments and demands for change.

Keywords Ideal and non-ideal theories · Dynamic obligations · COVID 19 pandemic · Vaccine global allocation · Patent waiver policy · WHO's international agreement

5.1 Introduction

The last two chapters focused on vaccine distribution during the pandemic and presented the empirical reality as well as the inadequate solutions implemented by the various actors (countries, international organisations, etc.). A faint possibility for improvement emerged towards the end of the previous chapter: COVAX Plus (or a similar mechanism) points to a needful future path. These modifications are possible and must be made to what was actually implemented.

Up to this point we have mainly worked with the decision-making mechanism by evaluating the different responses provided. This chapter goes further by taking up the concept of dynamic obligations to help us move towards the fifth requirement: considering long-term objectives. As discussed in Chap. 1, these obligations serve to expand our ability to meet certain principles or ideals over time. Unlike normal duties, dynamic duties do not focus on achieving certain desirable outcomes within the prevailing circumstances. Instead, they aim to change those circumstances so as to

F. Luna, *The Ideal and Non-Ideal in Bioethics*, The International Library of Bioethics 115, https://doi.org/10.1007/978-3-032-14594-9_5

make desirable outcomes achievable, often by expanding an agent's power (Gilabert 2017).

This concept is fundamental to exercising our political imagination and moral deliberation. And in this case, dynamic duties should serve to rethink elements for the next pandemic. How might we conceive of a change of circumstances? Which elements are essential to help us take this step? In this chapter we explore the possibility of changing reality to allow for the implementation of new instruments such as a patent waives or new international agreements.

5.2 What About Patents? Some Significant Concepts and Facts[1]

Limiting nationalisms and complementing them with mechanisms such as COVAX Plus remains an unsatisfactory approach. COVAX Plus would work even better if it could be refined with the introduction of dynamic obligations that change the prevailing circumstances. One of these could be waiving patents during the pandemic, a demand that many countries and institutions made.

A patent is an exclusive right granted by the administrative authorities of a state to the creator of an invention to exploit it economically for a certain period of time (usually twenty years) and within that territory. In the pharmaceutical field, patents protect products, such as medicines and vaccines, and the processes required to make them. Thus, the laboratories that obtain patents are the only ones entitled to market these products exclusively in the country where they obtained the patent. This legal monopoly allows patentees to set the price of the product and decide how to market it, which can make large-scale production at affordable prices difficult (Terlizzi 2021: 112). As Sol Terlizzi points out, "New drugs, ventilators and their parts, and vaccines can be patented in most countries. New uses of existing drugs and diagnostic kits are not patentable in countries such as Argentina. While personal protective equipment (PPE: gowns, face masks, etc.) is usually in the public domain, though some process or part may be patented" (Terlizzi 2021: 112, translation mine).

There are also different legal ways to avoid patents, such as compulsory licences. These are authorisations granted by governments to laboratories to produce or use a patented product or process without the authorisation of the patent holder, issued under certain conditions (temporary, payment of remuneration, etc.) and for certain purposes linked to public health, non-exploitation or non-production, national defence or development. Non-commercial public use is a mechanism that allows the patented invention to be used by the state, but not commercially. These tools are legitimate instruments that states can use in their public policies. However, they are not easy or quick to implement. Licensing takes time and may be subject to judicial

[1] Special thanks go to Sol Terlizzi for her careful reading, advice and suggestions for this and the next section.

review. In addition, it is difficult to identify all patents or other intellectual property rights involved in a vaccine or medicine. Compulsory licences apply only to granted patents, not to pending patents. They are granted preferentially to supply the domestic market, not for export; this makes them useless in countries with no technological-pharmaceutical development[2] (Terlizzi 2023).

A *waiver* can also be proposed. This is a permit granted by World Trade Organisation (WTO) members that authorises a WTO member not to comply with normal commitments.[3] Exemptions have to be time-limited and extensions have to be justified. This approach was proposed during the pandemic: a campaign to lift patents temporarily, #FreePatents, was initiated early on by India and South Africa.[4] It was backed by more than 100 countries along with international organisations such as WHO and UNAIDS, NGOs such as Médecins Sans Frontières and others (Editorial, *Nature* 2021).

In October 2020, India and South Africa made a formal request for a waiver of patents and other intellectual property rights. The proposal stated that certain provisions of the Agreement on Trade-Related Aspects of Intellectual Property Rights (TRIPS)[5] would not apply to any product or procedure to prevent, contain or treat COVID-19 until widespread vaccination had taken place worldwide, and the majority of the world's population had developed immunity.[6] Although it was debated for eight months, the US, the EU, the UK, Switzerland, Canada, Japan and Brazil rejected this request (Terlizzi 2021: 116).

In a change of course, the US[7] backed the lifting of vaccine patents (Editorial, *Nature* 2021). On 5 May 2021, the Biden administration in the US surprised the world with the announcement that it would support the lifting of patents (Iacobucci 2021). This attempt was blocked by Japan, South Korea, the UK and the European Union (especially Germany)[8] (*Nature* 2021 Editorial). Note that the WHO estimated that by 5 May 2021 (one of the peak times when all countries were desperately seeking

---

[2] For a detailed analysis of each of these arguments, see CEIDIE (2022: 38–41).

[3] It is found in Article IX.3 of the Marrakesh Agreement, which established the WTO: "In exceptional circumstances, the Ministerial Conference may decide to exempt a Member from an obligation imposed by this Agreement or by any of the Multilateral Trade Agreements, provided that such a decision is taken by a three-fourths majority".

[4] Regardless of the wisdom of the argument, economic interests may also be suspected here, as both India and South Africa have companies producing generic medicines.

[5] The TRIPS Agreement of the World Trade Organisation (WTO) is one of the most comprehensive multilateral intellectual property agreements. It allows its members to provide comprehensive intellectual property protections and was signed in Marrakesh in 1994.

[6] WTO—IP/C/W/669—Waiver of Certain Provisions of the TRIPS Agreement for the Prevention, Containment and Treatment of Covid-19 Communication from India and South Africa, 2 October 2020. https://docs.wto.org/dol2fe/Pages/SS/directdoc.aspx?filename=s:/IP/C/W669.pdf&Open=True.

[7] Statement by Ambassador Katherine Tai (2021).

[8] "German Chancellor opposes release of COVID vaccine patents". *El financiero*. 7 May 2021. https://www.elfinanciero.com.mx/mundo/2021/05/07/canciller-alemana-se-opone-a-liberacion-de-patentes-de-vacunas-covid/.

vaccines) 80% of the 1.1 billion doses had gone to high-income countries[9] (e.g. most of the US population had already been vaccinated) with only 3% administered to low-income countries. Although support from the US, Russia and China was significant, systematic opposition from the European Union meant that no progress was made by 2021. In an effort to unblock the negotiations, the QUAD group, made up of the United States, the European Union, India and South Africa, presented a proposal for an agreement to the WTO in March 2022. The proposal was approved with modifications on 17 June 2022, and the waiver to relax patents on coronavirus vaccines only rendered the patent system ineffective for a period of five years. The measure was intended to facilitate the production of vaccines by developing countries, and their eventual commercialisation in other low-income countries. While important, the proposal was unfortunately adopted too late, when the crisis had already happened and the pandemic was in its final stages. Furthermore, its scope was limited to coronavirus vaccines and does not cover, for example, any future pandemic (irrespective of the virus causing it).

The reluctance to provide a patent exemption is striking, as lifting patents during a pandemic or in public health emergencies was already an established policy position in international treaties but was unfortunately—as always—ignored. Although compulsory licensing and other mechanisms exist, they are very difficult to implement due to the legal and commercial stratagems that laboratories use to avoid them. The role of the pharmaceutical industry and producer countries in protecting their industry was key (Karp 2021).

5.2.1 What About Patents? Arguments Raised

If one looks at the various arguments put forward by representatives of pharmaceutical companies, one can see the reluctance of these actors to file patents. The classic argument that not respecting property rights undermines the incentive for companies to innovate was a recurring theme in the discussions, although it was not explicitly raised. This reasoning is undoubtedly indefensible at a time of pandemic, when research to alleviate the global crisis was often supported with public funds and especially when provisional expiry dates were proposed.

At the time, the International Federation of Pharmaceutical Manufacturers and Associations argued that lifting patents was not the solution to the crisis, and that in practice the big bottleneck in access to vaccines lay in the complexity of production (Editorial, *Nature* 2021). Some of the detractors of the usefulness of lifting patents made different arguments: it was likely to lead to disruption, while distracting from addressing the real challenges in scaling up production and distribution of COVID-19 vaccines globally, namely, elimination of trade barriers, addressing bottlenecks in supply chains and scarcity of raw materials and ingredients in the supply chain, and a willingness by rich countries to start sharing doses with poor countries (Iacobucci

[9] High-income countries, according to the definitions by the World Bank.

2021). The Chief Executive of the Bioindustry Association also noted: "Just handing countries' governments a recipe book without the ingredients, safeguards, infrastructure, and sizeable workforce with the high skills needed to deliver safe and effective vaccines will not speedily deliver help to all those that need it" (Iacobucci 2021).

Producing vaccines is much more complex than producing drugs or medicines, with minor changes altering the purity, safety and efficacy of the final product. Not only the final product, but also each stage of production and the place where production takes place are licensed (Bollyky and Bown 2020). However, the levels of difficulty vary depending on the type of vaccine, with more traditional viral vector vaccines such as the Oxford-AstraZeneca vaccine being more accessible because other vaccines are manufactured in a similar way and several sites already produce them, while mRNA vaccines such as Pfizer's and Moderna's require much more sophisticated sites and processes—although these are not necessarily better because they require longer cold chains, which makes them difficult to supply to some countries or places with less infrastructure. If appropriate training had been started in 2020, many of the arguments against the non-patent exemption fall apart, as several countries would have had good or at least better production conditions and could have started with viral vector vaccines (Holzer et al. 2021). So Iacobucci's arguments cited above do not hold. Training, including establishing training requirements in contracts with pharmaceutical companies, would have been one of the major tasks to be carried out by COVAX and is a requirement for COVAX Plus. In fact, as already explained, Argentina, Mexico and Brazil already have some infrastructure that should have been improved. In Argentina, a public–private venture is developing the ARVAC-Cecilia Grierson vaccine to protect against the variants circulating in the region.[10]

But perhaps more telling is the concern that qualifying other sites might result in increased competition The Director General of the European Federation of Pharmaceutical Manufacturers and Associations stated: "A waiver risks diverting raw materials and supplies away from well established, effective supply chains to less efficient manufacturing sites where productivity and quality may be an issue" (Iacobucci 2021). However, if there is adequate training, adequate qualification, and good products are achieved, the problem that really seems to concern this person is that qualifying other sites may result in competition to what already exists... Of course this adds to the diminishing returns for the pharmaceutical companies involved and for the producing countries. Bear in mind that Europe and the US (along with India) were the main producers of vaccines and made juicy profits.

Unfortunately, it was not only the pharmaceutical industry that adopted this position, but some well-known bioethicists also raised objections to the lifting of patents with very similar arguments (Emanuel et al. 2021). In their article "What are the obligations of pharmaceutical companies in a global health emergency?", these authors take the issue of "voluntarily" lifting patents and sharing know-how and secrets as a possible approach to obligations for pharmaceutical companies but criticise it as

[10] By mid-2023, it had completed two-thirds of the corresponding clinical trials. It is being developed by the National University of San Martin, CONICET and the Cassará laboratory (see: https://www.argentina.gob.ar/noticias/vacuna-arvac-comienza-el-ultimo-tramo-de-fase-23-y-convocan-).

unsustainable. They also argue that it would lead to a shift in investment (Emanuel et al. 2021: 1017; Persad 2018). And they point out that this strategy points to the wrong bottleneck as it is the knowledge or know-how of production and capacity that is limiting and not patents (Emanuel et al. 2021: 1016).[11]

5.2.2 Some Arguments in Favour

On the other hand, those who support the provisional lifting of patents offer different arguments. Terlizzi (2021) argues that denying the lifting of intellectual property rights in this situation goes against the original spirit and objectives of this type of regulation. Intellectual property rights, and particularly patents, were established to stimulate the creation, technological innovation, transfer and dissemination of technology, as stated in Article 7 of the TRIPS Agreement.[12] Intellectual property rights are a tool, a means to achieve various objectives such as scientific and technological development and improving human health.[13]

Such rights were established for normal times, not pandemic times. In the particular case of the pandemic, a commitment to distribute more equitably can be made, not only on grounds of distributive justice or the right to health, but also as a reciprocity argument. Much of the vaccine research for COVID-19 was carried out with public funding: Johnson & Johnson received $2 billion, Moderna $5.36 billion, Pfizer (BioNTech) $5.95 billion and AstraZeneca (Oxford) $1.6 billion (Petersen Institute of Economic Studies, in Ragavan 2021). Thus, publicly funded innovation should not be strictly a matter of private appropriation. A right to claim exemptions can then be formulated on the basis of that initial funding: the greater the public funding, the fairer it is for firms to share in the results they arrive at using such funds (Ragavan 2021). From the point of view of states, then, it can be argued that the greater the funding, the stronger their right—and even obligation—to request exceptional and forceful measures such as the *waiver* of such patents. In the face of such arguments, funding states can respond—as they did de facto—by protecting the interests of their companies and therefore of their citizens with a purely nationalistic view and by

[11] Although I had originally been a party to this article, I refused at the time to sign it because of this lukewarm position.

[12] "Article 7. Objectives. The protection and enforcement of intellectual property rights shall contribute to the promotion of technological innovation and to the transfer and dissemination of technology, to the mutual advantage of producers and users of technological knowledge and in a manner conducive to social and economic welfare and to a balance of rights and obligations". TRIPS Agreement (art. 7).

[13] "Article 8. Principles. 1. Members may, in formulating or amending their laws and regulations, adopt measures necessary to protect public health and nutrition, or to promote the public interest in sectors of vital importance to their socio-economic and technological development, provided that such measures are consistent with the provisions of this Agreement. 2. Appropriate measures may be necessary, provided that they are consistent with the provisions of this Agreement". TRIPS Agreement (art. 8).

not allowing the desired waiver. This is what can be inferred, for example, by the German reluctance to agree to petitions on patent waivers.

Secondly, the lifting of patents would imply legal certainty: that is, being able to produce vaccines in other countries with all the legal safeguards.

Thirdly, if used effectively, it would allow for the increase of local capacities that would not only benefit the economic development of other countries (those that are not the usual producers) but would also form a base structure for dealing with future pandemics. In all these cases, the patent exemption is provisional, is linked to the duration of the pandemic and is only justified in a public health crisis or emergency situation.

Finally, as we will discuss below, an incentive framework could be established to consider voluntary licensing and other benefits as part of an international regulation (Ortiz Millán 2024).

If dynamic duties are incorporated in this framework, it changes the prevailing circumstances. Patent sharing would contribute to moving towards a more ideal world; but even from a non-ideal (transitional) perspective it could also be seen as a transformative path or a way to overcome certain injustices (Laurence 2021). Having a well-oiled mechanism in place to implement such exemptions quickly is critical. It raises the possibility of more regional diversity amongst producer countries with legal certainty and lower costs. Of course, such a proposal is partial, but it is an important step in the right direction. To be truly effective it must be complemented by regional development and capacity building, and a commitment to equitable distribution, as argued for in COVAX Plus.

5.3 International Regulation and Preliminary Discussions

From the perspective of dynamic obligations, it is no less important to secure a strong and serious international agreement that establishes global guidelines to be implemented at the time of the next pandemic. In an attempt to learn from the failure of what happened, in December 2021 the World Health Assembly issued a resolution to begin negotiations for an international instrument on pandemic preparedness. To this end, it established an Intergovernmental Negotiating Body (INB) to draft it. The aim was to adopt it at the WHO World Health Assembly on 27 May 2024. At the 2025 World Health Assembly, while this book was about to be sent to the publisher, this agreement was finally approved.

Substantive issues were debated, ranging from early detection and prevention of outbreaks of infection and early warning systems, to universal access to medicines, vaccines and diagnostics. Broader issues included those advocating for One Health.[14] The Independent Panel on Pandemic Preparedness and Response

[14] The Food and Agriculture Organisation of the United Nations (FAO), the Office International des Epizooties (OIE), the United Nations Environment Programme (UNEP) and WHO are working on the integration of the One Health approach. This is an integrated unifying approach that

(IPPPR) recommended that the treaty should consider mechanisms for funding pandemic preparedness and response, development and research, technology transfer and capacity building, and strengthening global health security obligations and standards (Wenham et al. 2022: 6). The WHO and a group of heads of state noted that such a document should lead to greater mutual accountability, shared responsibility, transparency and cooperation with the international system and its rules and norms (Joint Statement by Head of States and WHO 2021). Note that several of these recommendations such as technology transfer and capacity building were raised as pending debts in COVAX and are called for in my COVAX Plus proposal.

At the WHO General Assembly, many states argued for the treaty to be developed in terms of human rights, solidarity and equity, including redressing the failures that occurred during COVID-19.[15] Indeed, the cosmopolitan and global spirit of the original proposals was permeated by the values of solidarity and equity, and the recommendations were received very positively by the academic world.[16]

But there were also sceptical and cautious views that highlighted the extremely nationalistic and self-interested real-world behaviour of most countries, particularly, but not exclusively, high-income countries. There is a danger of a profound erosion of the initial proposals and desires, and some authors have questioned the extent to which this international treaty project is a distraction from the current governance failure (Wenham et al. 2022: 16).

Creating a harmonious international legal regime presented many challenges (Wenham et al. 2022: 9). The treaty faced issues of governance and of international politics and diplomacy, such as whether it should be a WHO instrument ratified by states but represented by the health ministers of each country, or whether it should be at the level of the UN General Assembly, whose representatives are the heads of governments, in order to have greater force and be respected (Hannon et al. 2022). One of the problems of progressing a treaty is that consensus can be reached at the level of the highest principles but then one has to work with different protocols, which not only take a long time to implement but also have to be ratified by different countries, and such ratifications have in the past generated reluctance (Wenham et al. 2022: 9). For example, the US has ratified almost no human rights treaties. Another alternative is to combine it in some way with the International Health Regulations (IHR) that are already in place.[17]

seeks to balance and optimise the health of people, animals and ecosystems in a sustainable way. This involves considering health, food, water, energy and the environment in an integrated manner. See https://www.who.int/es/news/item/01-12-2021-tripartite-and-unep-support-ohhlep-s-definition-of-one-health.

[15] For example, Portugal and the Netherlands spoke about human rights, Chile, Norway, Ecuador, Egypt, Algeria, among others, about solidarity, and more than 60 countries, including the United Kingdom, Gabon, Georgia, Tanzania and the Dominican Republic, spoke about equity (see note 33 in Wenham et al. 2022).

[16] Among many works are Gostin et al. (2022), Burci et al. (2021), Vinuales et al. (2021), Frieden and Buissonniere (2021) and Khor and Heymann (2021).

[17] For example, there was a debate as to which articles of the WHO Constitution (19 or 21) it should be framed under (Ravelo 2022).

Compliance with these instruments is another highly complex point. Economic incentives to help improve infrastructures in different countries when there is timely reporting or adequate epidemiological surveillance is one possibility. Sanctions are another, but they can be counterproductive, for example, a country might hide an outbreak for fear of being sanctioned.

At the time of writing, the treaty was in negotiation. Initially, it was proposed to modify the existing WHO mechanism. The IHR, which were introduced more than 50 years ago, were amended in 2005, and adapted again in 2024, based on the experience of the pandemic. However, they were not respected during the pandemic (Hannon et al. 2022; Wenham et al. 2022). Another possibility (which is the one that was finally agreed) was to establish a legally binding treaty.

5.3.1 First Sketches

Nine draft versions of this treaty were published between 2022 and 2024. The texts submitted during 2023 demonstrate general trends. In February 2023 the WHO published what is known as the Zero Draft (WHO 2023a) of the pandemic treaty. It stated that this instrument was needed because of "the catastrophic failure of the international community in showing solidarity and equity in response to the coronavirus disease (COVID-19) pandemic" (WHO, Zero Draft 2023a). The goal, among other things, was to prevent pandemics, save lives and reduce the burden of disease, guided by equity, vision, principles and established rights, and to respond comprehensively and effectively to the systemic gaps and challenges that exist in these areas at national, regional and international levels (WHO, Zero Draft 2023a: 10).

On 2 June 2023, a revised version of this document was published, entitled "Bureau's text of the WHO convention, agreement or other international instrument on pandemic prevention, preparedness and response" (WHO 2023b) or, in its short form, WHO CA+. It weakened several provisions, but also offered formulations and language options for the future treaty. A comparison of the two versions reveals differences even in the title and order of the articles. In the Zero Draft there is a clear intention to promote equity and improve capacities in general. Words such as "promote" sustainable and equitably distributed technology production and transfer (Article 7), and "increase" research and development capacities (Article 9) are used in the titles. Two articles in Chapter IV explicitly "strengthen and sustain" capacities for health systems to prevent, prepare for, respond to and recover from a pandemic. These significant terms implying positive actions for the achievement of the proposed objectives do not feature in the titles of the later version.

Of the many topics covered by the treaty, there are at least three key strategic points that we will discuss further:

(1) how patents and intellectual property of drugs and products needed to alleviate the pandemic are dealt with;

(2) how it is financed and/or who will have access to the products to be distributed in a pandemic; and
(3) the role and responsibility of states with greater resources and corporations such as pharmaceutical companies.

The second version removes the introduction and states that it will be taken up later by the INB. However, the introduction to the Zero Draft was both revealing and eloquent. For example, it addressed the first key point mentioned: intellectual property (IP). Of the 49 paragraphs that made up the introduction, ten paragraphs in a row deal with IP, and while the importance of IP in the development of new medicines is highlighted (&40–42), the impact on pricing, timely production and distribution is also noted (&41–42). For example, paragraph 40 calls for enhancing the global effort towards innovative options for the production of, timely and equitable access to, and distribution of health technologies and know-how, including through local production. It recognises the concerns that intellectual property raises, the challenges and barriers to the realisation of the right to health and scientific progress of life-saving medical technologies (the price effect that limits access to options and impedes independent local production) as well as the need to establish some mechanism for pandemic prevention that goes beyond the charity model and appeals to justice. Finally, paragraphs 44 and 45 point to the flexibility and safeguards of the TRIPS Agreement. None of these points appear in the second version. However, in the June 2023 version, the issue of intellectual property is explicitly formulated in Article 11 under two alternative proposals. Option A states, among other points, "take appropriate measures to support time-bound waivers of intellectual property rights" (Article 11 option A 5.a) and "apply the full use of the flexibilities provided in the Agreement on Trade-Related Aspects of Intellectual Property Rights (TRIPS Agreement)" (Article 11 option A 5.b). In contrast, option B proposes to suspend the application of property rights through temporary lifting where "Neither Party shall challenge these measures based on any international obligations that the Party suspending the obligation may have" (Article 11 option A 5.e); an alternative formulation provides for deleting the latter paragraph (Article 11, option B 5.e) and leaving everything to the goodwill of the Parties (Article 11 option B 3.a and 4). Furthermore, it states that "The Parties [may]/[shall, as they deem appropriate,] make use of the full flexibilities provided in the TRIPS Agreement" (Article 11 option B 6). Khosla, McCoy and Marriot also refer to all these options and the dangers they entail (Khosla et al. 2023: 2019–2020): not only can all the options in option A be avoided, but one of the alternatives is that it can be removed in order to incorporate a public health perspective. Thus, following option B would leave us in a worse position than we are currently.

Other works support this critical reading. The *Lancet*'s editorial of 11 April 2023, significantly titled "The WHO pandemic treaty: promises of equity must be kept", discusses the second key point (funding and access to commodities) and notes that "The strongest commitment to equity is Article 10, which states that WHO should have access to 20% of 'pandemic-related products', so that it can distribute them to priority populations identified in national health plans. Half of this 20% should be provided to WHO free of charge, and the remaining half should be provided at 'affordable prices'; however, the document does not detail who should be providing these products" (*Lancet* Editorial 2023). Nevertheless, the editorial is optimistic and welcomes this first draft as a "genuinely positive step towards a more equitable pandemic response". Unfortunately, this optimism is not maintained when the second version of 2 June 2023 is published. In the article "Renouncing human rights and equity in the pandemic agreement", Rajat Khosla, David McCoy and Anna Marriot point out how the right to health and equity has been weakened (see also People's Vaccine Alliance 2023). The authors point out with regard to Article 10 that there is a possibility that the treaty includes language that binds pharmaceutical companies' access to pathogens or genomic data if they give up 20% of new products (10% as donations and the other 10% at affordable prices) for distribution in developing countries; the new formulation of the text also removes all obligations and makes developing countries totally dependent on voluntary corporate action.

Khosla, McCoy and Marriot also highlight the third key element: the role and responsibility of resource-rich states and corporations. The Zero Draft was designed to hold pharmaceutical companies accountable but Article 10 of the new version weakens this. "Provisions for pharmaceutical companies to commit to improving access to medicines, providing transparent public reporting of prices of pharmaceutical products, and ensuring equitable data sharing and technology transfer and know-how for production of pandemic-related products as condition for receiving public funds for research and development have been diluted" (Khosla et al. 2023: 2019). They continue: "Instead of imposing legally binding obligations and requirements on the corporate sector to act in the public interest and for the common good, the latest draft of the Pandemic Accord seems to be almost entirely reliant on voluntary acts or so-called corporate social responsibility" (Khosla et al. 2023: 2019). And this is raised with publicly funded research. The only duty bearers are now states, and other key actors are not included. These authors also argue that Article 14 on human rights has been removed and its implementation is absent or invisible. They argue for a reassessment of the appeal to human rights, as happened in the response to HIV-AIDS, to place limits on the abuse of power by state and non-state actors, redressing the imbalances of groups with powerful commercial interests (Enoch and Piot 2017).

In a similar vein, Ortiz Millán raises the need to seriously discuss the moral obligations of pharmaceutical companies, which should not be seen merely as economic entities, nor should vaccines be seen as 'commodities' (Ortiz Millán 2024). Another critical position along similar lines is that of Lavery et al. (2023).

5.3.2 *What About Principles?*

Both documents begin by proposing different principles to help to achieve and implement the treaty's objectives. The original eighteen principles are reduced to twelve in the second version. The first principle of the June 2023 document, respect for human rights, includes in its wording the right to the highest attainable standard of health—the right to health corresponding to the second principle of the Zero Draft that has now been removed and subsumed into this human rights principle. The principle of equity, formerly the fourth principle and now the third, is better drafted as it begins by putting equity at the centre of the response. But, suspiciously, it does not state that there is a need to "tackle inequities in technology, infrastructure, financing" as the previous formulation did. The only principle that is actually better than in the Zero Draft version is the principle of transparency, because in addition to what was previously presented, it now includes reporting on sales, revenues, prices, units sold, marketing costs, subsidies and incentives.

In their analysis of the proposed principles, Schaefer and colleagues (2023) focus on the principles themselves, their different status and importance, rather than on the possible weakening introduced by the second version and its impact on the principles. They begin by properly pointing out the importance of incorporating ethical principles. They recognise that the twelve principles that remain are not all ethical and that this can lead to confusion. Though perhaps true, even if some of them are not considered ethical, they may nevertheless be relevant to the way the treaty operates and translates.

Schaefer et al. propose eight ethical principles that are taken from other WHO documents (WHO 2016, 2021b). These are

(1) maximise benefits and minimise harms,
(2) equity,
(3) global solidarity,
(4) sustainability,
(5) accountability,
(6) transparency,
(7) engagement, and
(8) science and evidence-informed decisions.

Schaefer et al. (2023: 2) distinguish between substantive and procedural principles: substantive principles indicate the right decision or optimal outcome and correspond to their first four, whereas procedural principles (the other four) relate to processes and how signatories should make decisions. This distinction, in my view, also makes it possible to give priority to substantive principles.

A review of this proposal shows that the authors incorporate two new principles. One is to maximise benefits and minimise harms: they define it as "to protect and promote human well-being including physical and mental health, social and economic security, human rights, and child development, while minimising the harmful effects of policies (whether direct or indirect, health-related or not)" (Schaefer et al. 2023: 3).

The other new principle is that of sustainability, defined as: "Ensure that emergency responses that are appealing for the immediate problem do not imperil future responses, preparation for the next pandemic, or responding with research, development and manufacturing to subsequent pandemics" (Schaefer et al. 2023: 3). Unlike

Schaefer et al., however, I consider this a procedural principle. While it is important that certain principles can be sustained in the long run (see requirement 5 of our decision-making mechanism to choose between ideal and non-ideal theories in Chap. 1), it points more towards the operationalisation of the substantive principles which are those that should govern fundamental decisions.

I agree even less with the analysis and application of the latter principle to the discussion on intellectual property and patents. There, Schaefer et al. raise doubts about the sustainability of the provisional lifting of patents and refer to empirical issues for decision-making. Given the exorbitant profits that the pharmaceutical companies involved made during the pandemic, the same fairness and solidarity demanded of all actors should also be strongly applied to them. This situation does not seem to be present in the work of these authors.[18]

Schaefer et al.'s proposed eight principles leave out the principle of respect for human rights (principle 1) and the principle of sovereignty (principle 2), as they argue that these are implicit in a treaty of this nature. In a similar spirit they remove the principle postulating the central role of the WHO (principle 11), which they have taken for granted. Similarly, they do not include the principle of inclusiveness (8), the principle of proportionality (12) and the one-health principle (9). However, another more relevant principle, which Schaefer et al. eliminate, is the "principle of differentiated capabilities and responsibilities".

This principle is highly relevant in the dynamics of our globalised world. It is evident in climate change treaties—most notably in the United Nations Framework Convention on Climate Change (UNFCCC 1994)—and has gained broad recognition in international law (Stone 2004; Cullet 2010). Upon closer examination, we see that while the term 'common' refers to risks that affect all nations and require global cooperation, the responsibilities are differentiated: not all countries contribute equally to global warming (in the case of climate change), nor do they possess comparable resources for building responsive health systems or surveillance infrastructures (as in pandemic preparedness). In the latter case, some countries are better positioned to contribute—whether financially, technologically or through expertise. This principle allows for the flexibility to recognise differing capacities among states: for instance, poorer member states may require assistance and collaboration. At the same time, it may be justifiable to prioritise countries that have contributed public funds to research on pandemic treatments or vaccines, particularly in terms of timely access to resulting products. Such prioritisation can help avoid hoarding or redundant procurement processes, such as occurred during the COVID-19 pandemic. These kinds of collaborative and constructive practices can serve as incentives for ethical global action. Although the principle of common but differentiated responsibilities and capabilities is not frequently invoked, it is highly pertinent in light of the pandemic—especially considering that some countries lack the economic means or expertise necessary for adequate disease surveillance.

[18] Something similar appears in the article by Emanuel et al. (2021), which I refrained from signing because of what I regarded as the overly condescending attitude towards the pharmaceutical industry. See the section on patents in this chapter.

The principle of differentiated capacities and responsible undeniably involves normative content with an attempt to seek greater equity and justice. It corresponds to the third key point: the role and responsibility of resource-rich states and corporations. It was present in the Introduction (paragraphs 32 and 36) as a concept and it appeared in the Zero Draft as the eighth principle. In the second version it is stipulated as the seventh principle, but again, three formulations are proposed. Thus, option 7A is called "common but differentiated responsibilities and capacities". It points out, in line with the Zero Draft, the uneven development in different countries and that those with greater capacities and resources should bear a proportionate degree of differentiated responsibility for pandemic prevention, preparedness and response. This option incorporates the notion of equity and maintains the original spirit of the Zero Draft. The second proposal (7B) seeks to avoid the problem by noting that the uneven development of countries is a "common danger". No obligation or response is stated here. In other words, it merely provides a description of the situation. And the third version (7C) directly proposes to eliminate the principle. Thus, if options 7B or 7C prevail, the possibility of states with greater resources having greater responsibilities is eliminated. Schaefer et al. do not take this principle into account and do not even address it.

Unfortunately, trends suggest that the principle of differentiated capabilities and responsibilities is disappearing. The April 2024 version raises it as a fifth principle but obscures and blurs it by referring to solidarity and a mere recognition of different levels of capacities and capabilities.

5.3.3 Reviewing Strategies

What can we make of all these discussions? In the article by Schaefer et al. we find interesting proposals, some ordering of principles, but also some questionable interpretations and omissions. The criticisms of those who follow this process closely are highly relevant (Khosla et al. 2023; Heidari et al. 2023; Burki 2023; Hanbali et al. 2023). Even INB members themselves are aware of the many challenges facing the drafting of this instrument (da Silva Nunes et al. 2023). For example, this agreement, instead of being debated by technicians and specialists on the issues at stake, is being discussed by diplomats and country representatives with their self-interest in mind, and continues to be influenced by the powerful pharmaceutical industry.

The question is: will this treaty with its annexes and new mechanisms be up to the task? While it was less ambitious than expected, many commentators agree that it retains its core objective: preventing future pandemics and ensuring that, should they occur, the world responds in a more equitable and coordinated manner at a global level. However, this is only the first step, and critical issues remain unresolved. An annex still needs to be drafted and negotiated to establish a Pathogen Access and Benefit-Sharing System (PABS) as well as a Coordinating Financial Mechanism for pandemic prevention, preparedness and response, along with the Global Supply Chain and Logistics Network (GSCL Network). Once the Assembly adopts the PABS

Annex, the agreement will be open for signature and potential ratification by member states. It will enter into force once at least 60 countries have ratified it. After that, its effectiveness will depend on each country's implementation—meaning several steps still lie ahead.

Two of the most significant challenges are: first, to achieve a relevant treaty complete with annexes and financial coordination mechanisms that effectively establishes adequate criteria to prevent pandemics, save lives and reduce the burden of disease in ways that respect equity. Second, if a pandemic arises and such a treaty will have to be effectively implemented, it will be the most powerful countries that must set limits on themselves and their pharmaceutical companies, but as we have just seen, these same countries have been unwilling to submit to such constraints, and there is a risk that the terms or limits will be ambiguous or too lax (Borges et al. 2022). Another way of framing these difficulties is to point to the prevalence of nationalist or statist actions versus the neglected cosmopolitan or globalist view (Wenham et al. 2022). And this is perhaps why it is most appropriate to approach it within the framework of dynamic obligations. As previously mentioned in Chapter Two, political imagination must be used to envision alternative ways of fulfilling certain principles. Dynamic duties direct a change, often an expansion of an agent's power to achieve certain outcomes, including establishing obligations that allow for a way out of this non-ideal world.

I believe that we must not throw in the towel when facing the sceptical view that relies on the status quo and the failures that existed during the pandemic, where nationalistic and statist behaviour reared their ugly heads. This is precisely where we must appeal to ideal theory and dynamic obligations.[19] Now is the time to bring about change and to try to reverse behaviour that is known not to work. We have the opportunity to modify certain circumstances precisely so that the results we desire can be achieved. We can see how a substantive international treaty that supports a more equitable world and is also legally binding fulfils the five requirements of the decision mechanism proposed in Chap. 1.

With regard to the first requirement—that it be politically feasible—this would seem to be the ideal moment, as there is both interest and incentive to break from the past and resist lobbies, pressures or weakening demands. However, this would not be easy or straightforward. As it remains as one of the challenges. If the proposal, along with its amendments and mechanisms, were to incorporate the most relevant ideals put forward to ensure cooperative and equitable functioning, it would satisfy the third requirement (i.e., being morally permissible). The fourth requirement would call for prioritising the most egregious injustices, as it would aim to assist the least advantaged nations in addressing their plight. A well-drafted instrument would also fulfil the fifth requirement, since it should embody equitable objectives to move beyond a Hobbesian state of nature and achieve more harmonious, long-term global cooperation. For instance, when considering patent waivers, it might stipulate that they be limited to pandemic-related needs and include temporary or provisional measures to

[19] Or the non-ideal, for example, as a transitional theory approach in which dynamic obligations function as agents of change. See Laurence (2021).

prevent undesirable long-term effects. Thus, rather than creating obstacles, a robust international treaty could lay the foundation for a promising future. Moreover, by pre-establishing cooperative rules, a legally binding instrument could circumvent the prisoner's dilemma. Once again, incorporating dynamic obligations would reshape reality. The most challenging requirement to meet would likely be the second one—effectiveness. This is where the second major challenge arises: determining when to implement the treaty during a pandemic and how different states would behave. Overcoming this would require combating scepticism, inertia and entrenched interests in the status quo.

Perhaps to provide a better basis for this proposal for change, let us think back to post-World War II. Facing the horror of what had happened, the United Nations system was built, different human rights treaties were designed, and international institutions such as the WHO were created with the aim of laying the foundations for a better global coexistence. After a global pandemic that stopped practically the entire world from circulating, closed borders, generated unseen poverty, collapsed health systems and caused so many deaths may also be the historic moment to establish other guidelines for our global coexistence, complementing what has already been done to achieve true equity. A well thought-out and properly designed mechanism of this type is fundamental to change the panorama and the rules of the game.

5.4 Conclusion

We began with the problem of how to moderate nationalisms and move them towards less extreme positions. Faced with the Hobbesian situation posed at the beginning of the third chapter, one possibility is to veer towards a moderate nationalism as in the flu-risk standard explored in that Chapter. But it has been seen that this is a minimal solution that does not solve the problem. It does allow country to be accountable to its own residents and to emerge from the crisis and emergency by respecting national sovereignty, but setting limits and aiming for solidarity-based behaviour beyond its own borders. Not only is it far from the ideal, it is unfortunately also far from what actually happened.

Secondly, COVAX was a promising first step that changed historical behaviours, but also had obvious shortcomings. COVAX Plus brings in substantial and relatively easy modifications and improvements. If this proposal is to be acceptable, the Fair Priority Model (FPM) must first be adopted as a basic criterion of equity so that all countries are considered on an equal basis, regardless of their income or resource levels. To achieve diversity and some regional independence, it will be important to work with COVAX Plus or a similar mechanism on training and vaccine development in regional centres in developing countries, enabling them to move away from dependence on the few countries that currently produce vaccines. And finally, to solve the problems of transparency and efficiency, COVAX Plus could very well make all these changes, which, moreover, are in line with its mandate and objectives.

COVAX Plus may serve to alleviate some of the effects of the pandemic, but above all to prepare the ground for dynamic obligations such as the provisional lifting of patents. This would make it possible to lower costs during a pandemic and, subject to regional training, to produce vaccines or medicines in a more diversified and regional manner. Finally, a strong international body promoting collaboration and fairness is needed to regulate and set limits for countries and pharmaceutical companies. To achieve this, an international pandemic treaty with a strong legal mandate has been considered. For Gostin, Klock and Halabi this is the most transformative global health call to action since the WHO itself was formed as the first specialised agency of the United Nations in 1948 (Gostin et al. 2022).

We are facing a historic moment in which this dynamic obligation can be deployed to change and modify old patterns that are becoming increasingly obsolete, especially in the light of pressing environmental issues and global warming. Increasingly, the fact that we live in a single world with a single health system is being pointed out. The international outlook, despite the extreme nationalisms of the real world, invites us to defend a supportive vision. This is the trend towards which the world must turn in the face of sceptics who point out that these proposals seem doomed to failure. More cautious positions point out that it may not happen quickly or in the medium term, but with awareness-raising campaigns, pressure from civil society, etc., dynamic changes and obligations such as those outlined above could be put in place. This is one of the great challenges that we as a society must face.

We can begin with small changes, such as those proposed in the previous chapters, taking into account the weight of this non-ideal world. But we must also look beyond: trying to reach the island paradise and start designing the tools that will point us in the right direction and allow us to achieve that goal. Of course, we will face tides and storms, but the time has come to build new tools, ships with the right instruments that will bring us closer to the longed-for island. And if we manage to establish a world in which greater trust and respect prevail, perhaps we will also be able to escape from the sinister prisoner's dilemma.

References

Bollyky, T., and C. Bown. 2020. The tragedy of vaccine nationalism. *Foreign Affairs*, September–October.

Borges, L. C., H. Z. de Menezes, and E. Crosbie. 2022. More pain, more gain! The delivery of COVID-19 vaccines and the pharamaceutical industry's role in widening the access gap. *International Journal of Health Policy Management (IJHPM)* 11 (12): 3101–3113.

Burci, G. et al. 2021. *Envisioning an international normative framework for pandemic preparedness and response: issues, instruments and options*, 401, 1255. Graduate Institute Geneva.

Burki, T. 2023. Calls for transparency in Pandemic Accord talks. *The Lancet* 401 (10384): 1255–1255.

CEIDIE. 2022. *El debate respecto al pedido de exención de los derechos de propiedad intelectual en el Consejo de los ADPIC & la propuesta de entendimiento*. CEIDIE.

Cullet, P. 2010. Common but differentiated responsibilities. In *Research handbook on international environmental law*, ed. M. Fitzmaurice, D. M. Ong and P. Merkouris.

da Silva Nunes, T. et al. 2023. Addressing inequity: The world needs an ambitious Pandemic Accord. *The Lancet* 402: 271–273.
Emanuel, E. J., et al. 2021. What are the obligations of pharmaceutical companies in a global health emergency? *Lancet* 398: 1015–1020.
Enoch, J., and P. Piot. 2017. Human rights in the fourth decade of the HIV/AIDS response: An inspiring legacy and urgent imperative. *Health and Human Rights* 19: 117–122.
Frieden, T., and M. Buissonniere. 2021. Will a global preparedness treaty help or hinder pandemic preparedness? *BMJ Global Health* 6: 5.
Gilabert, P. 2017. Justice and feasibility. A dynamic approach. In *Political utopias. Contemporary debates*, ed. M. Weber and K Vallier, 95–126. Oxford University Press.
Gostin, L. et al. 2022. Inching closer to an essential global pandemic treaty. STAT, First Opinion, August 3. https://www.statnews.com/2022/08/03/inching-closer-to-an-essential-global-pandemic-treaty/.
Hanbali, L., et al. 2023. Independent monitoring for the Pandemic Accord: A non-negotiable provision. *The Lancet* 401: 553.
Hannon, E., et al. 2022. Why we still need a pandemic treaty. *The Lancet* 10 (9): E1232–E1233.
Heidari, S., et al. 2023. A gender responsive Pandemic Accord is needed for a healthier, equitable future. *The Lancet* 402 (10418): 2176–2179.
Holzer, F., et al. 2021. A matter of priority: Equitable access to COVID-19 vaccines. *Swiss Medical Weekly* 151: w20488.
Iacobucci, G. 2021. COVID-19: How will a waiver on vaccine patents affect global supply? *BMJ* 373 (11821).
Karp, P. 2021. Pfizer warns Australia a COVID vaccine patent waiver could harm supply and safety. *The Guardian*, 20 de enero. https://www.theguardian.com/australia-news/2021/may/13/pfizer-warns-australia-a-covid-vaccine-intellectual-property-patent-waiver-could-harm-supply-and-safety.
Khor, S., and D. Heymann. 2021. Pandemic preparedness in the 21st century: Which way forward? *Lancet Public Health* 6 (6): 357–358.
Khosla, R., et al. 2023. Backsliding on human rights and equity in the Pandemic Accord. *Lancet* 401: 2019–2021.
Lancet. 2023. WHO's pandemic treaty: Promises of equity should be kept. *Lancet*, 11 (e475).
Laurence, B. 2021. *Agents of change*. Harvard University Press.
Lavery, J. V., R. M. Porter, and D. O. Addiss. 2023. Cascading failures in COVID-19 vaccine equity. *Science* 380: 6644.
Naciones Unidas, Convención Marco de las Naciones Unidas sobre Cambio Climático (1994).
Nature. 2021. A patent waiver on COVID vaccines is right and fair. *Nature* 593 (478).
Ortiz Millán, G. 2024. Ethical and practical considerations for an agreement to ensure equitable vaccine access. *International Journal of Health Policy Management (IJHPM), 13*, 8516.
Persad, G. 2018. Examining pharmaceutical exceptionalism: Intellectual property, practical expediency and global health. *Yale Journal Health, Policy Law Ethics* 18: 157.
Ragavan, S. 2021. Waiver IP Rights & Save Lives. *SouthViews* 231, 29 November. https://www.southcentre.int/wp-content/uploads/2021/11/SV231_211129.pdf.
Ravelo, J. 2022. Majority of WHO member states want legally binding pandemic instrument. *Devex*, 21 July. https://www.devex.com/news/majority-of-who-member-states-want-legally-binding-pandemic-instrument-103669.
Schaefer, G., et al. 2023. The importance of getting the ethics right in a pandemic treaty. *The Lancet Infectious Diseases*, July 5. https://doi.org/10.1016/S1473-3099(23)00364-X.
Stone, C. D. 2004. 'Common but differentiated responsibilities' in international law. *American Journal of International Law* 98 (2): 276–301.
Tai, K. 2021. Statement on the COVID 19 Trips Waiver. United States Trade Representative, March 5, en https://ustr.gov/about-us/policy-offices/press-office/press-releases/2021/may/statement-a,bassadpr-katherine-tai-covid-19-trips-waiver.

Terlizzi, M. 2021. La pandemia patentada. In *Bordes. Revista de Política, Derecho & Sociedad.* Universidad Nacional de José C. Paz.

Terlizzi, M. 2023. Clase: Pandemia y patentes; un drama en tres actos. Maestría en Bioética, FLACSO.

Vinuales, J. et al. 2021. A global pandemic treaty should aim for deep prevention. *Lancet* 397m: 10287.

Wenham, C., et al. 2022. The futility of the pandemic treaty: Caught between globalism and statism. *International Affairs* 98: 837–852.

World Health Organization (WHO). 2016. *Guidance for managing ethical issues in infectious diseases outbreaks.* The World Health Organization. https://iris.who.int/handle/10665/250580

World Health Organization (WHO). 2021a. Director-General's opening remarks at 148th session of the executive board. The World Health Organization, 18 January. https://www.who.int/director-general/speeches/detail/who-director-general-s-opening-remarks-at-148th-session-of-the-executive-board.

World Health Organization (WHO), WHO access to COVID-19 Tools Accelerator Ethics & Governance Working Group. 2021b. *Ethical framework for WHO work in the ACT Accelerator.* The World Health Organization, July.

World Health Organization (WHO) & head of States. 2021. *Joint Statement: COVID 19 shows why united action is needed for more robust international health architecture.* The World Health Organization, March 30. https://www.who.int/news-room/commentaries/detail/op-ed-covid-19-shows-why-united-action-is-needed-for-more-robust-international-health-architecture.

World Health Organization (WHO). 2023a. *Zero Draft of the WHO CA + for the consideration of the Intergovernmental Negotiating Body at its fourth meeting. WHO convention, agreement or other international instrument on pandemic prevention, preparedness and response (WHO CA+).* The World Health Organization, February 1.

World Health Organization (WHO). 2023b. *Bureau's text of the WHO convention, agreement or other international instrument on pandemic prevention, preparedness and response (WHO CA+).* The World Health Organization, June 2.

World Trade Organization (WTO—IP/C/W/669). 2020. Waiver of certain provisions of the TRIPS agreement for the prevention, containment and treatment of Covid-19 communication from India and South Africa, 2 October. https://docs.wto.org/dol2fe/Pages/SS/directdoc.aspx?filename=s:/IP/C/W669.pdf&Open=True

Part III
Other Non-Ideal Conceptions

Chapter 6
Vulnerability and Its Theoretical Analysis

Abstract The third part of the book focusses on problems present in our real, everyday world. It works on another way of analysing non-ideality in key concepts for bioethics. It abandons the usual and traditional view to explore them and takes into account their non-ideality. It focusses primarily on the concept of vulnerability and then on the concept of relational autonomy. The sixth chapter sets out how the concept of vulnerability should be analysed. It criticizes the traditional conceptualization, for example, its rigid analysis and assumptions. It proposes to think of vulnerability using the metaphor of layers. It then points out the dispositional nature of the layers of vulnerability and highlights the importance of identifying the stimulus conditions or possible elements that trigger such vulnerabilities. It also introduces the notion of cascading vulnerability that can generate incremental damage. It is also explained how this analysis includes other definitions of vulnerability, and the relations with the feminist concept of intersectionality. The central core of this part shows how our way of conceptualising the concept of vulnerability affects our resolution of it.

Keywords Vulnerability · Theoretical analysis · Layers of vulnerability · Stimulus conditions · Cascade Vulnerability · Definitions of vulnerability · Intersectionality

6.1 Introduction

In Part I, I presented two cases that required seemingly different responses, before focusing on the scope of ideal and non-ideal theories. In Part II, I focused on our non-ideal world through the lens of global vaccine distribution in pandemics, as well as looking at possible ways out through the ideal. I applied the decision mechanism formulated in Chap. 1 to determine how acceptable or unacceptable the different answers provided were, as well as proposing some ways to move towards the ideal through dynamic duties. I now shift focus from ideal theory to concepts used in

This is a concept that I have worked on extensively and on which I have taken a position of my own. This part is based on previous articles: Luna (2009a, b), Luna and Vanderpoel (2013); Luna (2014, 2015, 2019, 2022); and Buedo and Luna (2021).

F. Luna, *The Ideal and Non-Ideal in Bioethics*, The International Library of Bioethics 115, https://doi.org/10.1007/978-3-032-14594-9_6

our non-ideal world. In Part III I consider other ways of analysing key concepts for bioethics, with a particular reference to non-ideality.

I move beyond simplistic approaches, for example, interpretations of the principle-based theory (Beauchamp and Childress 1979) frequently reduce bioethics to a rigid 'cookie-cutter' application of four principles, a sort of checklist that magically solves every dilemma or problem. The theory of principles became the gold standard in bioethics. Early versions of Principles of Biomedical Ethics (Beauchamp and Childress 1979) (up to and including the third) of this deductivist model were based on four principles, but it was then contested and adapted to different criticisms, and a coherentist model and a so-called common-sense ethics was defended. Although requiring significant theoretical modifications, it always kept the principles at its core. Dubbed 'principlism', the popular use of this theory oversimplified bioethics by applying principles dogmatically to all cases. In addition to the highly problematic difficulties of interpretation, the theory was modified in different editions of the book Principles of Biomedical Ethics (Beauchamp and Childress 1979), but the changes that were introduced were never explicitly clarified, and above all, those who read the book were never alerted to the differences that were established (for example, the same name was kept for the theory and for the different editions without even including explanatory subtitles, which leads the reader to think that it is the same initial approach).[1] However, while bioethics does of course use certain principles, it is not exhausted by them. Deeper and broader approaches are needed; there must be a deliberative analysis in which key concepts are examined and incorporated into our reflections. As we shall see, the concept of vulnerability has a leading role in this discipline.

This chapter considers in depth how the concept of vulnerability should be analysed,[2] closely related as it is to the context. It can also be read through a non-ideal lens. I use the metaphor of layers rather than the traditional label model to elucidate it. I point out the dispositional nature of the layers of vulnerability to highlight the importance of identifying the stimulus-condition and I introduce the notion of cascading vulnerability. These two features of vulnerabilities have an impact on the practical analysis that I will then propose in the next chapter. I also show how my analysis includes other definitions of vulnerability. Other definitions of vulnerability include the notion of cascading vulnerability. Chapter 7 will propose an analysis that responds to theoretical-conceptual problems but also brings substantial practical content. The theoretical strategy proposed plunges us into the non-ideal world, showing the strength of the layers of vulnerability and their potential for real harm, but its translation into practice offers concrete guidelines for dealing with them. The greatest merit of this conception of vulnerability is that it does not simplify or idealise research participants, research contexts or public health problems. Rather, it accepts the challenge of providing a non-simplified response to a complex problem.

---

[1] For more detail on this theory and its problems see "Planteos clásicos y teoría de los principios (Classical approaches and principled theory)" in Luna (2008: 23–78).

[2] I have been working on this concept in its different dimensions for more than a decade: see Luna (2009a, b), Luna and Vanderpoel (2013, 2014, 2015, 2019, 2022), and Buedo and Luna (2021).

Chapter 8 briefly considers another concept that can be thought about in non-ideal terms—that of autonomy, especially relational autonomy and its analysis from a feminist perspective. The concept of relational autonomy moves away from the initial principled approach and the traditional ethical model. It recovers the non-ideality present in this concept and broadens our way of thinking about bioethics.

6.2 The Conceptual Sphere: Traditional Analysis

The term 'vulnerability' is key to research and public health ethics. The concepts of 'vulnerability' and 'vulnerable groups' have been widely used in these fields, but were not critically examined until well into the twenty-first century. It was not until 2004, with the publication by Carole Levine, Ruth Faden, Cristine Grady, Dale Hammerschmidt, Lisa Eckenwiler and Jeremy Sugarman of the article "The Limitations of 'Vulnerability' as a Protection for Human Research Participants" that this concept began to be seriously examined by bioethicists.

Previously, interesting analyses had mostly gone unnoticed: from the philosophical analyses of Robert Goodin (1986) to the more specific analyses of Kenneth Kipnis (2001) commissioned by the US National Bioethics Advisory Commission on the occasion of the drafting of the Belmont Report. Perhaps the examination of fundamental and critical concepts was not considered due to the impact of the theory of principles, which enjoyed a great boom at the beginning of this discipline and which seemed to exhaust all possible analysis in bioethics. In 2008 Samia Hurst published the article "Vulnerability in Research and Health Care; Describing the elephant in the room?" and my own article "Elucidating the Concept of Vulnerability: Layers not Labels" was published in 2009. Hurst and I addressed this challenge directly, proposing substantive solutions. Since then, the field has become increasingly aware of the vital role of this concept, which could not be subsumed under the uses made of it in research ethics documents. Different, controversial reflections were put forward.

In the article "*La declaración de UNESCO y la vulnerabilidad, la importancia de la metáfora de las capas* (The UNESCO declaration and vulnerability, the importance of the metaphor of layers)", I drew a distinction between two spheres of discussion (Luna 2009b: 258–9). The first can be conceived as an ethical-conceptual sphere which analyses the theoretical and normative problems of the conceptualisation of vulnerability. The second covers the ethical-practical sphere—the concrete practice required to protect certain people and to maintain the relevance and usefulness of the concept of vulnerability for the evaluation of protocols or for the formulation of public policies. Here I use the terms 'conceptual sphere' and 'practical sphere'.[3]

Vulnerability, then, has an important normative role as well as practical and political force. Two fundamental questions capture these challenges:

1. How should vulnerability be understood (conceptual question)?

---

[3] In the original article I called them "theoretical-ethical" and "political-ethical" but I think this new terminology is clearer.

2. How should it specifically be used (practical question)?

These two spheres of reflection are closely related, for if we cannot provide an adequate response to the conceptual sphere, we cannot defend its practical relevance. In other words, if this concept is vacuous or stereotypes populations, it cannot be successfully applied. These serious conceptual problems must be resolved in order to restore its usefulness and normative force. At the same time, the two spheres pose different problems. While the conceptual sphere focuses on the correct analysis and understanding of the concept of vulnerability, the practical develops how the concept can be applied to the ethical evaluation of research and the design of public policies.

Why has this concept generated so much interest? Since the Belmont Report was published in the 1970s, there has been a tendency to label particular sub-populations as vulnerable. I have called this approach "the metaphor of labels" (Luna 2009a: 122–3). Labelling sub-populations as 'vulnerable' oversimplifies complex situations, ignoring research complexity and context, and the varied vulnerabilities experienced by individuals and groups. Hence a non-ideal lens is needed to analyse this concept.

In 2004 Carole Levine, along with other well-known bioethicists, criticised the broad use of the concept of vulnerability as too vague to be meaningful (Levine et al. 2004), pointing out that it generated stereotypes: "the concept of vulnerability stereotypes whole categories of individuals, without distinguishing between individuals in the group who indeed might have special characteristics that need to be taken into account and those who do not" (Levine et al. 2004: 47). These authors are right to point out this problem. First of all, stereotyping implies putting a label on people or groups that cannot be easily removed. And secondly, stereotyping or labelling people can harm or damage them. It also prevents us from identifying different levels of vulnerability. The concept is presented in a dichotomous way: you are either included or excluded. But the reality is that among vulnerable people, some may be worse off than others. This lack of flexibility leads to a rigid view of the situation and implies conceptual and normative problems (Luna 2009a: 127–8).

Another strong criticism pointed out by these bioethicists is that there are now so many categories of people who can be considered vulnerable that "*virtually all potential human subjects are included*" (Levine et al. 2004: 46, emphasis mine). They cite the CIOMS 2002 version of the guidelines, among other documents, arguing: "*Under one or another of these rubrics, nearly everyone is vulnerable,* especially since the benefits of research can never be guaranteed in advance …. *If everyone is vulnerable, then the concept becomes too nebulous to be meaningful*" (Levine et al. 2004: 46, emphasis mine).

The overpopulation of vulnerable groups implied by Levine et al.'s criticism is exacerbated by multi-centred international research that conducts clinical trials in developing countries where so-called vulnerable populations abound. This suggests that there are essential characteristics that populations or people meet: one is either included or excluded in a categorical way. But this approach is problematic in that it implies general moral obligations for all, but does not infer special moral obligations for particular situations. If this broad approach were to prevail, it would justify the need for general protection of human beings in research (a point that research ethics per se makes).[4] However, when the notion of vulnerability in research is used, what is sought is to provide special or specific protections or safeguards for particular persons that cannot be achieved from the general perspective. Moreover, this view ends up 'naturalising' vulnerability.

These deep problems at the ethical-conceptual level of analysis (vacuous use, stereotyping effect, etc.) have important consequences at the practical level (Levine et al. 2004: 127–8). These criticisms need, in the first place, a response from the heart of the conceptual sphere through a non-ideal lens. Levine et al. are right to point out some of these problems mainly in relation to the conceptual question. There is an inadequate consideration of the "reality" of the context and its impact in this discussion. However, on the practical level, the critics' position is rather restrictive. Though they do not say so explicitly, Levine et al. reject this concept, accepting regulations only for research with children (Levine et al., 2004: 44 and 47), and special protections for people with permanent cognitive impairments (Levine et al., 2004: 48). Note they are adopting the traditional concept of autonomy with the limits this implies. Levine et al. (2004: 44, 47–8) adopt the traditional concept of autonomy with the limits this implies, accepting regulation only for research with children and special protections for people with permanent cognitive impairments.[5] They also propose "special scrutiny" as an alternative way of providing specific protection for participants.[6] The image of 'throwing out the baby with the bathwater' seems to aptly illustrate the resolution proposed by these authors. While the criticisms of Levine et al. that I consider valid are those directed at the conceptual sphere of analysis, the 'baby' must be recovered. That is, a proper analysis of the concept of vulnerability must be provided in order to be able to apply it correctly. Vulnerability remains highly relevant in practice, and in failing to fully address the first sphere these authors have discarded it or made it overly narrow.

6.3 Is Vulnerability an Ontological or a Relational Property?

It is often questioned whether vulnerability should be understood as an ontological or relational property. As long ago as 1996, Onora O'Neill made a very clear distinction between two conceptions of vulnerability. She pointed out that the first was related to us as a species. Vulnerability is an expression of the human condition, of its fragility and finite nature, and is different from being vulnerable because of specific circumstances. The second conception explains that an individual or group of individuals can become "deeply, variably and selectively vulnerable to the action

---

[4] For example, when protocols are required to be evaluated by a research ethics committee or when informed consent is required, etc.

[5] This point will be further developed in Chap. 8.

[6] See also Luna (2009a: 128).

of the particular others and the particular institutions on whom we come to depend for specific and often for unavoidable purposes" (O'Neill 1996: 192).

One response points out that vulnerability is an essential and universal property of humanity, inherent to all people. This kind of thinking has a long tradition in continental philosophy (Ricoeur 2007; Levinas 1968) but has also had a very strong impact on feminist thinking. For feminism with an ontological approach, people are fundamentally and essentially vulnerable (Butler 2004; Callahan 2000; Fineman 2008; Mackenzie 2014; Mackenzie 2016). Butler explores the ethical implications of the inherent vulnerability of the human body. She explains that 'precariousness' (our vulnerability to the actions of others) generates obligations to minimise suffering and eliminate inequalities, but also points out that some individuals and populations are disproportionately precarious. Note the tension that this last point generates: on the one hand, precarity affects all people, but on the other hand, there are some people who are disproportionately precarious! Thus the ontological approach does not allow us to distinguish between one form of precariousness and another.

Martha Fineman (2008, 2010) is another feminist philosopher who considers vulnerability as something "universal and constant, inherent to the human condition" (Fineman 2008: 1). She dissociates this concept from negative connotations such as neediness and victimhood, while emphasising it as an ontological condition of our embodied humanity. She uses this notion of corporeal vulnerability in the context of a critique of liberal legal and political theory. She proposes a theoretical perspective for understanding inequality and disadvantage through the concept of the vulnerable subject. Other feminists, while sympathising with this approach, point out that there is a problem with understanding vulnerability as something opposed to autonomy (Mackenzie 2014: 33). For Mackenzie, an adequate ethics of vulnerability must give a central place to the promotion of autonomy since focusing only on protections (the traditional way of conceiving vulnerability) can generate paternalisms and coercive forms of intervention.[7]

The opposing view sees vulnerability as situational and relational (Goodin 1986; Macklin 2003; Luna 2009a, 2019; Armstrong 2017). For example, for Goodin, vulnerability essentially means being under threat of harm and arises from particular relationships (Goodin 1986: 110). Similarly, for Macklin, a person is vulnerable if conditions are such that they are at greater than normal risk of being harmed (Macklin 2003). Here again, the context is introduced. This approach allows specific policies or safeguards to be designed for these individuals or groups. We do not all fall equally under the veil of vulnerability.[8]

Frédéric Armstrong's attempt to resolve this dichotomy is particularly interesting: this Canadian author analyses the strategy of Rogers et al. (2012) and their taxonomy as a possible response and then that of Martin et al. (2014) to criticise and discard them and sustain a relational proposal. In fact, Armstrong relies on the 'layering' proposal as the correct way to settle this dispute and recover the relational proposal.

---

[7] For more detail on the subject see Luna (2022), Rogers (2017).

[8] See also Luna (2022).

While one could accept an anthropological analysis in which the concept of vulnerability is understood ontologically, the use of the term vulnerability in research ethics or public health ethics is related to the second dimension distinguished by O'Neill, that is, the relational one. There is also a confusing approach in the critiques of Levine et al. On the one hand, they seem to be pointing to an overpopulation of vulnerable groups, but their way of expressing this fact seems to lead to the idea that vulnerability is a general trait or characteristic that all human beings share.

This kind of ontological position, moreover, carries the risk of 'naturalising' vulnerability. But if we are all vulnerable and this is a "natural fact", an expression of human nature, as O'Neill would say, that we all share,[9] then there is no need to avoid it or to protect certain people (Luna 2009a: 128). In this case, the general protections of research ethics would suffice, and there would be no special cases or situations. This assumption is dangerous, especially in the current international research context, this would mean ignoring vulnerabilities in order to reject the problems of protecting or taking into consideration certain populations. It may also lead to avoiding obstacles to the conduct of research and any kind of testing could be permissible. Such logic is, at the least, cynical, if not dangerous. Denying the validity of the idea of vulnerability or blurring it as an ontological property may be more comfortable than acknowledging it and seriously providing safeguards to avoid harmful consequences.

It is perhaps more interesting to abandon the categories of relational or ontological and think of vulnerability in terms of a contingent property which someone or something may or may not have.[10]

6.4 The Structure and Functioning of Vulnerability: The Metaphor of Layers

In my article "Elucidating the concept of vulnerability", I proposed that the concept be conceived through the notion of *layers* (Luna 2009a: 121–139). The layers metaphor suggests multiple vulnerabilities that can be added or removed individually. There is no "one solid, single vulnerability", but there may be different vulnerabilities, like different layers operating. These layers may overlap: some may relate to informed consent, others to human rights violations, social circumstances, the context or the characteristics of the person involved. The layered approach has strong ties with the context and with the non-ideal vision.

For example, we might say that being a *woman* does not in itself make one vulnerable. But a woman living in a country that does not recognise or is intolerant of *reproductive rights* acquires a layer of vulnerability (as opposed to a woman living in a country that does respect reproductive rights). In turn, an educated woman with

[9] This is not to say that O'Neill holds this position, as she distinguishes between the two and specifies when one or the other should be used.

[10] I thank Eleonora Orlando and Diego Lawler for this suggestion.

economic and social resources in the same country may be able to overcome some of the consequences of non-recognition of reproductive rights. However, a *low-income woman* living in a country that does not respect reproductive rights acquires another layer of vulnerability: she may not have access, for example, to emergency contraception and will therefore be more susceptible to unwanted pregnancies. Moreover, a low-income, *illiterate* woman in a country that does not recognise reproductive rights acquires yet another layer. And if she is a *migrant* and does not have her documents in order and does not speak the language or if she belongs to Indigenous peoples, she will acquire more and more layers of vulnerability. She will suffer from the sum and interaction of these layers (Luna 2009a: 127–128).

This concept of vulnerability is dynamic and contextual. A person may no longer be considered vulnerable if the situation changes. For example, a French woman of reproductive age and lower-middle income may not be vulnerable in a research protocol if she becomes pregnant unintentionally (because in her country she has access to emergency contraception or to an abortion in a public hospital if she wishes). On the other hand, if she is in El Salvador (where legal abortion is not allowed for any reason), that same French woman in that same protocol may acquire a layer of vulnerability.

A further problem with the label metaphor is that it involves two assumptions. First, it assumes a baseline standard that implies a default *paradigmatic* human participant (a mature, moderately educated, clear-thinking, literate and self-sufficient person) and, second, it identifies the vulnerabilities of sub-populations in opposition to or defaulting from the paradigm. This model both idealises and simplifies research subjects: not only do such subjects not exist, but such a vision denigrates those who do not correspond to the idealised. Thus can the ideal—in this case, bad idealisations—make its way into non-ideal concepts. Here it is worth recalling Mills' criticisms of idealisations considered in Part I. Idealising human participants not only affects our view of them as research participants but also impacts how we think about informed consent. Somehow these people are assumed to function as 'contractors'. I return to this point towards the end of this chapter.

Furthermore, the existence of diverse human participants within groups challenges the idea of homogeneous groups sharing the category of vulnerability. The sub-population approach assumes the existence of necessary and sufficient conditions that populations must meet to be considered vulnerable. In contrast, the layers of vulnerability approach allows for differences or variations within the group to be addressed, and different types of safeguards or empowerment tools targeted at different characteristics to be considered.[11]

If there are no longer rigid labels or categories, then, there are no fixed sub-populations: people who might traditionally be classified as vulnerable may not be vulnerable in a given situation and vice versa. Most interestingly, this conceptual analysis allows us to move away from 'usual' or 'typical' stereotypes. For example,

---

[11] The metaphor of layers can work with individual people but also with sub-groups within a population. The important thing is to consider the different layers at play and not just take one characteristic to represent the group as a whole, either for an individual or for a sub-group.

we tend to consider low-income and pregnant people as vulnerable. However, in certain situations this may not be true. Argentine law, for example, allows low-income pregnant women and pregnant women to collect umbilical cord blood through a public bank free of charge, covering the whole process in case a cord blood transplant is needed for a brother or sister. So, in this case, low-income pregnant women are adequately treated and covered. However, before recent regulations were introduced, commercial umbilical cord blood banks used to exert very strong pressure on middle-class pregnant women. They used, among other strategies, aggressive marketing and misleading information in private gynaecology practices, as well as manipulating the pregnant woman's feelings of guilt for not being a 'good mother' if she did not collect the umbilical cord blood of her future baby. With this type of public policy, it was the middle-class person who could most readily acquire a layer of vulnerability (Luna and Vanderpoel 2013: 325–332). In recent years, these practices have been regulated to avoid this kind of subtle coercion. But this example demonstrates how those who in principle would have been categorised as vulnerable were not, and how pregnant women from better socio-economic conditions—often not considered vulnerable—acquired a layer of vulnerability.

The sub-population approach and the label metaphor are inadequate because they amount to using the concept of vulnerability as a mere slogan, categorising and stereotyping people. The layers of vulnerability approach, in contrast, allows us to include all the variants and situations that people face in our non-ideal world without simplifying any one element. My approach proposes that vulnerability:

1. has no single standard or ideal, but recognises multiple factors or sources of vulnerability;
2. is deeply context related; and
3. is not an essential property of individuals or research groups per se.

6.5 The Meaning of Vulnerability: Definitions and Content

The concept of vulnerability is slippery and elusive. In order to understand it better, a distinction must be made between:

(a) the structure and functioning of the concept, and
(b) its 'content' or characterisation.

What has just been made explicit in the previous section relates to point a) which corresponds to the structure of the concept. Thus, its functioning is relational and dynamic, closely linked to context and the situation under analysis, rather than being a category or a label that we can simply apply. The layered approach 'unpacks' the concept of vulnerability and *shows how it works*.

The conceptual content (point b) accommodates multiple definitions. My layered proposal does not contradict the main conceptualisations or definitions of the concept of vulnerability; on the contrary, it can include them. This is the case, for example, with definitions of vulnerability such as those pointing to a lack of power and the possibility of exploitation (Zion et al. 2000: 615), the inability of research subjects to protect their interests (CIOMS), or risk—and harm-based approaches (Hurst 2008).

These different ways of defining vulnerability can be expressed through several coexisting layers, without there being a need to choose a single definition. For example, the inability of research subjects to protect their interests identified by the 2002 CIOMS definition can be translated into layers related to informed consent or into layers related to inadequate safeguards to protect research participants. Moreover, the 2016 version of the CIOMS guidelines retains the 2002 definition, but also includes my proposed layers (although it does not use the word 'layer') (CIOMS 2016).

There are also layers related to socio-economic conditions that could put people at greater risk or subject them to exploitation. Samia Hurst explains that "vulnerability as a claim to special protection should be understood as an identifiably increased likelihood of incurring additional or greater wrong" (Hurst 2008: 195). Hurst shows how the concept relates to other concepts such as wrongs, harms and normative claims. And again, the likelihood of harm can be translated into different layers. These definitions emphasise different layers highlighting multiple factors that express vulnerability, and we are not obliged to choose only one of them.

Ruth Macklin (2003: 69–70) suggested a different way of looking at this complementary approach. Macklin criticised Hurst's proposal for its narrowness.[12] She points out that such a proposal only applies to research and clinical care, and cannot easily include public health. Macklin argues that Hurst's notion of a "claim to special protection" is too narrow as it cannot account for circumstances in which culture, custom, tradition and law mean that, for example, women cannot protect themselves. Macklin reinterprets Hurst's analysis and considers that my approach can complement it, bringing a contextual and relational aspect to this definition.

Another compatible approach is Kenneth Kipnis' identification of some "characteristics that are criteria of vulnerability" (Kipnis 2001). He proposes a taxonomy with six circumstances: cognitive, juridical, deferential, medical, allocational and infrastructural.[13] In a subsequent article on the vulnerabilities of those involved in paediatric research, Kipnis adds a seventh category, social circumstances (Kipnis 2003). This addition jeopardises the idea of a clear and efficient taxonomy.[14] (I will expand on this point later). However, in Luna (2009a: 126 and 134) I stress that Kipnis'analysis goes in the same direction as mine. While Kipnis also seeks an analytical rather than a sub-population approach, he differs from my analysis in proposing a taxonomy. Despite this, I think we can follow Kipnis' analytical approach without committing ourselves to taxonomies. Each of these characteristics can serve as a guide. If taken as a kind of open-ended list, they can help to identify the layers, but they may be too narrow and rigid if we stick only to them as they are set out in a taxonomy.

---

[12] This was true at least at that point of Hurst's writings. However, she continued working with a team (Martin and Tavaglione), so some of her positions may have changed.

[13] See the section "Taxonomies and corsets" in Chap. 7.

[14] For more critiques of taxonomies, see Luna (2015).

6.6 Stimulus Conditions or Triggers

My own proposal accepts all these different definitions and conceptualisations, as they show the multiple angles of this layered approach to vulnerability. My early analyses are set out in Luna (2009a, b, 2014). Following some interesting criticisms that challenged my thinking, however, I continued working on this concept. These second developments led me to point out and analyse more clearly two features of the concept of vulnerability (Luna 2019).

The first feature is that layers of vulnerability are dispositions and that the structure and relevant characteristics of these dispositions must be taken into account in an ethical analysis. While a layer of vulnerability reveals that a person may be abused, mistreated or exploited in certain circumstances, it is not necessary for a person to be abused, mistreated or exploited for them to be considered to have a layer of vulnerability or to be vulnerable.[15] Moreover, if they are abused, mistreated or exploited, they are no longer vulnerable because they *have already been* harmed, abused, etc. In Spanish, there are two different words to point out this difference: *vulnerable*, which has this dispositional character and corresponds to the English term, and *vulnerado* which is no longer a disposition but an established fact: the person has been harmed or affected. It is the possibility of being harmed, abused or exploited that is relevant in the concept of vulnerability. Hurst's analysis also assumes this dispositional character when she speaks of an "*increased likelihood* of incurring additional or greater harm" (Hurst 2008: 195). The relevance of this aspect of the concept of vulnerability at the practical level, however, has not to my knowledge been shown.

The notion of disposition and its analysis is not new in philosophy, but it is new for bioethics. An analysis of the dispositional structure of layers of vulnerability can help us to reach a thorough ethical evaluation. A classic example of a disposition is the property of being soluble. A sugar cube has the property of being soluble. That is, it dissolves if it is immersed in water or a liquid, the key element here being the stimulus condition. The action of introducing the sugar cube into a liquid is the stimulus condition that will cause the sugar cube to go from being a solid to dissolving into many particles. Dispositions, then, are latent until triggered by a specific stimulus condition that actualises them. Just as we have identified the stimulus condition for the sugar cube to be soluble, so too must we identify the stimulus conditions that would trigger a particular layer of vulnerability. Identifying triggering conditions is crucial for the analysis and assessment of layers of vulnerability. We therefore need to differentiate between the disposition (the vulnerability) and possible stimulus conditions (the triggers or elements that can actualise or activate them).

In the case of people participating in research, there can be many different situations that can trigger a layer of vulnerability. If those stimulus conditions do not occur, that layer of vulnerability will not be activated. Somehow, what has happened in research ethics is that it has been assumed that research per se functions as a stimulus condition, and that for certain populations or individuals, the mere fact of

[15] I thank Eduardo Rivera López for this comment.

participating in it triggers vulnerabilities. But this is not necessarily the case. A more careful analysis identifies what situations might trigger certain vulnerabilities, and assesses whether the research as a whole will actually make the person acquire layers of vulnerability, or whether it depends on, for example, some particular elements, and it is these that need to be modified. If a stimulus condition does not occur or is not likely to occur, the disposition will not be actualised. Hence, if we can identify triggers (and thus eradicate, avoid or minimise it), we can probably prevent the actualisation of such a disposition and thus the harmful consequences of layers of vulnerability. Researchers, research ethics committees (RECs) and policy makers need to distinguish and identify the stimulus conditions or triggers that may precipitate a vulnerability layer, in order to design appropriate protections and safeguard mechanisms.

6.7 The Cascading Effect of Layers of Vulnerability

The second feature is that some layers exhibit cascading effects. While I argue against some of the points of Meek Lange and Dodds (2013) and Rogers et al. (2012), I believe that they have nevertheless made very valuable contributions. An important contribution of Rogers et al. relates to the concept they call the pathogenic source of vulnerability. They explian the concept that patogenic source of vulnerability.

> …may exacerbate existing vulnerabilities or generate new vulnerabilities. We refer to these as pathogenic vulnerabilities. There are a variety of sources of pathogenic vulnerability. Pathogenic vulnerability may be generated by morally dysfunctional interpersonal and social relationships characterised by disrespect, prejudice, or abuse, or by socio-political situations characterised by oppression, domination, repression, injustice, persecution, or political violence. For example, people with cognitive disabilities, who are occurrently vulnerable due to their care needs, are susceptible to pathogenic forms of vulnerability, such as sexual abuse by their carers. (Rogers et al. 2012: 25)

Thus, I distinguish between two ways of characterising such layers of vulnerability:

1. Its origin: its generation by morally dysfunctional interpersonal and social relationships (the position highlighted by Rogers et al. in the paragraph above);
2. Its effects: the consequences of this type of layer (a chain of events leading to harmful consequences).

My proposal emphasises the relevance of the second feature. The source need not always have a "pathogenic" character, as Rogers et al. (2012) propose. Instead, the importance of this distinction lies in the normative force implied by its harmful effects: a potential to "*exacerbate existing vulnerabilities or generate new vulnerabilities*". Thus, I argue that what is evidenced by Rogers et al.'s pathogenic source is a replication or consecutive unfolding of harmful effects: a cascade effect. While there may be occasions of cascade effects generated by morally dysfunctional relationships, it is the normative force of their effects that is the most relevant feature to consider. Rogers et al.'s notion of pathogenic source is very useful, although it should be reformulated in a broader and more neutral sense since the term ‘pathogenic’ is associated with illness and pathology. I therefore propose that we call these layers of vulnerability—following the second identified characteristic—*cascading layers of vulnerability* and that we seriously consider the devastating power they can have.

For example, in public health ethics, the lack of early diagnosis of a rare disease can be seen as a cascading vulnerability (Luna 2019: 450). Lack of diagnosis implies lack of knowledge, and the uncertainty and anguish, as well as mistreatment, suffered by the individual and their family, generates further layers of vulnerability. Without proper diagnosis, the disease may progress, making it impossible for the person to access effective treatment when finally diagnosed. Parents may make reproductive decisions without being aware of the high possibility of giving birth to another baby with the same problem, etc. This example demonstrates that abusive relationships or oppressive socio-political situations are not the only situations that trigger this cascading characteristic. A layer of vulnerability may arise because the person is confronted with a rare disease, and in this case there is no mistreatment or abuse.

This feature of vulnerability is very much related to the non-ideal aspect outlined in this book. Cascading layers can often be identified in the field of public health, as they are associated with pre-existing conditions and contextual conditions (lack of access to health services, lack of protective labour laws, etc.), and public health policies are key to eradicating or minimising their effects. But in some cases, this vulnerability with a cascading effect can be very difficult to eradicate, even with laws or public strategies. In the case of policy design for older people, we can see different layers of potential vulnerability (economic, communicational, emotional, cognitive, physical, etc.), some of which can be considered cascading vulnerabilities (Luna 2014: 184–186).

For example, let us consider this possible layer: unlike the traditional family model that lasted a lifetime and in which strong bonds were generally established between family members during decades of marriage,[16] in today’s families there are separations, several marriages, a succession of cohabitations, and the new people in these more recent ties may not have established quality relationships with the older people in the family. In some cases, new partners may not be very accommodating towards older family members, resulting in a layer of vulnerability for some of them. We can call this the relational layer: older people feel lonely, isolated and a burden to their families. If this relational layer occurs, it can also be seen as a cascading layer, as it can lead to psychological damage, such as depression, which at the same time can lead to a loss of appetite or mobility, which in turn is likely to engender frailty and physical illness, and other morbidities.

Cascading vulnerabilities illustrate the functioning of our non-ideal world. Not only is this relevant theoretically, but cascading vulnerabilities also play a key role in the process of assessing the strength or harmful effect of layers. It therefore represents a key distinction.

---

[16] I am assuming functional families, although this is not always the case.

6.8 Intersectionality

Before concluding with the theoretical aspects of my proposal on vulnerability, I would like to examine the relationship and links that this analysis may have with the concept of intersectionality that is so relevant in feminism and with the non-ideal. The concept of intersectionality is related to the cascading feature of vulnerabilities, and similarly to the non-ideal.

The concept of intersectionality was formulated by the African-American feminist lawyer Kimberle Crenshaw in 1989 in relation to a legal case. With this concept, she sought to highlight the multiple dimensions of oppression that black women workers suffered.[17] Structural intersectionality proposes that sufferings of women of colour and race in terms of domestic violence, rape, etc. are qualitatively different from those suffered by white women.[18] This approach is framed within one strand of feminism: black feminism (Crenshaw Williams 1989). It is relevant to note that Crenshaw's aim had been contextual and practical, to analyse specific legal omissions and inequalities. It was not originally intended to create a theory of domination (Viveros Vigoya 2016).

There are multiple views and perspectives in feminism, from those that consider the concept of intersectionality as the greatest conceptual contribution of feminism in the last 25 years (McCall 2005: 177; La Barbera 2016) to those who argue that intersectionality as a proposal is not new in feminism (Viveros Vigoya 2016).

Mara Viveros Vigoya traces this type of conceptualisation back to 1791, to Olympia de Gouges' declaration of women's rights in France. She compared colonial domination with patriarchal domination and drew analogies between women and slaves (De Gouges 1791). Viveros Vigoya takes up several manifestations of the same type of analysis in the USA and Latin America. For example, in Peru Clorinda Matto de Turner (1889) pointed out some of these intersections by revealing sexual abuses perpetrated by Governors and local priests on Indigenous women, highlighting how both ethno-racial and gender vulnerability played a role (Viveros Vigoya 2009). In the twentieth century there were several manifestations, such as those of the Combahee River Collective (1977–1983, 1986), Angela Davis (1981–2004), Audré Lorde (1984–2007), Bel Hooks (1984), Norma Alarcón (1990), Chela Sandoval (1991), María Lugones (2005, 2010), among many others, who spoke out against the hegemony of 'white' feminism because of the racial and gender biases of the category of 'woman' (Viveros Vigoya 2016: 4). Maria Caterina La Barbera (2016: 107) makes a similar point, noting that intersectionality is not an unprecedented concept and that it is the result of long debates on both sides of the Atlantic about the transversality of gender as an analytical category and essentialism as a danger for feminist theory. It should be noted that while the concept of intersectionality undoubtedly draws on and is the continuation of all these previous manifestations, it

[17] This is the case of General Motors (Crenshaw Williams 1989).

[18] I will use the term 'race' as it is the one that appears in all writings on the subject to highlight a particular form of discrimination, but it is a highly contested concept that is considered inadequate on scientific grounds.

crystallises, synthesises and operationalises these previous expressions, and in this sense introduces an extremely interesting conceptual leap.

What is the point of this new concept? Authors such as Audré Lorde (1982) or Angela Davis (1981) argued that by patterning the experiences of white, heterosexual, middle-class American women, feminism failed to understand how race, class and sexual orientation pluralise and particularise what it means to be a woman. Thus, the concept of intersectionality complicates previous analyses. Reconceptualising identity as multiple and contradictory shows that it is not fixed but an ongoing process of social construction (Harris 1990: 584). Interestingly, continuing in a similar spirit but using the categories of colonisation and diaspora, postcolonial feminists also warned of a particular discourse on Third World women in which history and social structures are not considered (Mohanty 1988: 80). Thus this concept captures the complexity of identity as it relates to many women, especially those from the peripheries.

There is also a double genealogy in the gestation of this concept: on the one hand, black studies (or black feminism) in the USA, and on the other, European postmodernism (Viveros Vigoya 2016). The former is linked to US law (as we saw in Crenshaw's proposal), while the latter is developed through the different social sciences: in sociology (Brah and Phoenix 2004; Bilge 2013; Ferree 2009; Mc Call 2005), in political science (Bustelo 2009; Hancock 2007; Kantola and y Nousiainen 2009; Lombardo and Verloo 2009), in philosophy (Garry 2011; Zack 2005) and in psychology (Greenwood and Aidan 2008). Thus for Kathy Davis, intersectionality is inscribed in the postmodern project of conceptualising identities as multiple and fluid and the Foucaultian perspective on power (Davis 2008: 71).

This concept has developed in different ways. In addition to that of political intervention, characteristic of feminism, discursive debates have arisen about the scope and content of intersectionality as a theoretical and methodological paradigm. There are also distinctions between macro-sociological and micro-sociological analyses. Maximalist approaches are proposed by Hill Collins (1990): "alternative paradigm to the antagonism between positivism and post-modernism which was part of the dichotomies structuring Western epistemology" (in Bilge 2009: 74). There are also analyses influenced by other traditions, such as those of Elsa Dorlin (2009). For her, there are two types of approaches to domination: the analytical one, which argues that by definition all domination is a domination of class, sex and race and is itself intersectional since gender cannot be coherently dissociated from race and sex; and the phenomenological approach, for which intersectionality is the experience of domination. Another critical element that is raised is the questioning of the hegemonic model of the universal 'Woman'. Geometrical formulations have also been made within intersectionality, versus arithmetical formulations of the addition or juxtaposition of identity elements. However, despite the fluidity that many of these analyses attribute to this concept, some theorists point out that this approach assumes the existence of groups and does not allow us to think about a changing and historical relationship of domination (Kergoat 2009). For this author, intersectionality implies fixed positions, and she resorts to the materialist feminist perspective to provide a better response (Viveros Vigoya 2016: 8).

Although the formulation of the concept of intersectionality can be traced back to 1989, we should bear in mind that in Latin America, this concept only began to be disseminated more broadly from 2008 on (Viveros Vigoya 2016: 7). Thus, if we focus on its theoretical developments, we can see that they are multiple and complex, drawing on different origins and philosophical traditions. As Viveros Vigoya points out, while some refer to gender, race and class as intersecting systems, others understand them as analogical categories or as multiple bases of oppression, and each of these approaches has its own theoretical implications (Viveros Vigoya 2016: 7).

Returning to the approaches to vulnerability, the concept of layers is not limited to women; it can also include men, boys or girls, groups with different capacities, complex socio-economic situations, and beyond. On the one hand, it shares with intersectionality the critical view of essentialist visions that tend to simplify situations and identities either through the label that such a population is vulnerable (because it supposedly deviates from the desirable 'research participant' model) as well as implying the standard of 'being a woman' corresponds to certain types of women (white, heterosexual, middle class). Thus, both conceptualisations imply a multiple, complex and dynamic view. On the other hand, the concept of vulnerability with a cascading effect seems to have similarities with the idea of intersectionality as something qualitatively different. It is not a mere sum of identities or, in our case, vulnerabilities. Vulnerability with a cascading effect is turbulent in its development and has harmful effects that multiply and trigger each other. In this sense, the two proposals speak the same language. Both analyses were developed in parallel and nourished by a similar critical vision, and in some ways they can perfectly dialogue and complement each other. Both seem to populate the non-ideal world that this book strives to illustrate.

6.9 Conclusion

In this chapter I address the theoretical and conceptual dimension of vulnerability, taking up again the points I made in 2008. There I pointed out that this concept should be understood through the metaphor of layers, showing that this type of analysis provides a contextual and flexible view and avoids labelling entire populations as vulnerable. This strategy makes it possible to recover the concept of vulnerability. Now I believe it is also related to non-ideal conceptualisations. I also analyse other proposals that naturalise the concept of vulnerability and take it as an ontological property that all people possess by virtue of the fact that they are people. I show that one of the great problems is precisely this risk of naturalising this concept, since in bioethics, and especially in research ethics or public health ethics, when reference is made to vulnerable people, the aim is to identify special conditions that make such people worthy of special protection or care, and not to 'humanity' or essential features of human beings. I then show how the metaphor of layers is related to the structure and functioning of the concept of vulnerability as distinct from its content or characterisation. In this sense, this conceptualisation of vulnerability accepts very different definitions emphasising its content.

My subsequent analysis of the dispositional structure of this concept highlights the existence of stimulus conditions or triggers that can precipitate vulnerability, and considers the domino or cascade effect of the features of certain layers, emphasising its non-ideality.

The chapter ends by examining the concept of intersectionality, which shares some elements with my layered proposal and the non-ideal perspective. Its significance in black feminist discourse is also considered. The different theoretical elements contained in the important but elusive concept of vulnerability are all used in the next chapter to reveal the applied aspect of the term.

References

Alarcón, N. 1990. The theoretical subject(S) of this bridge called my Back and Anglo-American feminism. In *Making face, making soul: Haciendo Caras. Creative and critical perspectives by women of color*, ed. G. Anzaldúa, 356–369. Aunt Lute Foundation.

Armstrong, F. 2017. An extrinsic dispositional account of vulnerability. *Les ateliers de l'éthique* 12 (2–3): 180–204.

Beauchamp, T., and J. Childress. 1979. *Principles of biomedical ethics*. Oxford University Press.

Bilge, S. 2009. Théorisations féministes de l'intersectionnalité. *Diogène* 225:70–88.

Bilge, S. 2013. Intersectionality undone. Saving Intersectionality from feminist Intersectionality studies. *Du Bois Review: Social Science Research on Race* 10:405–424.

Brah, A., and A. Phoenix. 2004. Ain't I a woman? Revisiting intersectionality. *Journal of International Women's Studies* 5 (3): 75–86.

Buedo, P., and F. Luna. 2021. Toma de decisiones compartidas en salud mental: una propuesta novedosa. *Medicina y ética* 32 (4): 1087–1110.

Bustelo, M. 2009. Spain: Intersectionality faces the strong gender norm. *International Feminist Journal of Politics* 11:530–546.

Butler, J. 2004. *Undoing Gender*. Routledge.

Callahan, D. 2000. The vulnerability of the human condition. *Bioethics and Biolaw* 2:115–122.

Council for International Organizations of Medical Sciences (CIOMS). 2002 (2016). International Ethical Guidelines for Health-Related Research Involving Human Subjects (3rd ed.) Geneva: CIOMS.

Combahee River Collective. 1977 [1983]. The Combahee River collective statement. In B. Smith (ed.). *HomeGirls, a black feminist anthology*. 272–282. Kitchen Table-Women of Colors Press.

Combahee River Collective. 1986. *Combahee river collective statement. Black feminist organizing in the seventies and eighties*. Albany: Women of Color Press.

Crenshaw Williams, K. 1989. *Demarginalizing the intersection of race and sex: A black feminist critique of Antidis-crimination doctrine, feminist theory and antiracist politics, 139*. University of Chicago Legal Forum.

Davis, A. 1981 [2004]. *Mujeres, raza y clase*. Akal.

Davis, K. 2008. Intersectionality as buzzword. A sociology of science perspective on what makes a feminist theorys uccessful. *Feminist Theory* 9 (67): 67–85.

De Gouges, O. (1791). *Declaración de los derechos de la mujer y de la ciudadana*. http://www.culturamas.es/blog/2012/09/07/declaracion-de-los-derechos-de-la-mujer-y-de-la-ciudadana-1791-por-olympe-de-gouges/

Dorlin, E. 2009. Introduction: Vers une épistémologie des résistances. In *Sexe, race, classe, pour une épistémologie de la domination*, ed. E. Dorlin, 5–20. PUF.
Ferree, M. 2009. Inequality, intersectionality and the politics of discourse. In *The discursive politics of gender equality: Stretching, bending and policy-making*, ed. E. Lombardo et al., 86–104. Routledge.
Fineman, M. 2008. The vulnerable subject: Anchoring equality in the human condition. *Yale Journal of Law and Feminism* 20 (1): 1–23.
Fineman, M. 2010. The vulnerable subject and the responsive state. *Emory Law Journal* 60:251–275.
Garry, A. 2011. Intersectionality, metaphors, and the multiplicity of gender. *Hypatia* 26:826–850.
Goodin, R. 1986. *Protecting the vulnerable: A re-analysis of our social responsibilities*. University of Chicago Press.
Greenwood, R., and C. Aidan. 2008. What happens when we unpack the invisible knapsack? Intersectional political consciousness and inter-group appraisals. *Sex Roles* 59:404–417.
Hancock, A. 2007. Intersectionality as a normative and empirical paradigm. *Politics and Gender* 3 (2): 248–254.
Harris, A. 1990. Race and essentialism in feminist legal theory. *Stanford Law Review* 42:581–616.
Hill Collins, P. 1990. *Black feminist thought. Knowledge, consciousness, and the politics of empowerment*. Routledge.
Hooks, B. 1984. *From margin to Center*. South End.
Hurst, S. 2008. Vulnerability in research and health care; describing the elephant in the room? *Bioethics* 22 (4): 191–202.
Kantola, J., and K. y Nousiainen. 2009. Institutionalizing intersectionality in Europe: Introducing the theme. *International Feminist Journal of Politics* 11:459–477.
Kergoat, D. 2009. Dynamique et consubstantialité des rapports sociaux. E. Dorlin (Ed.), Sexe, race, classe. Pour une épistémologie de la domination (pp. 111–126). París: PUF.
Kipnis, K. 2001. Vulnerability in research subjects: A bioethical taxonomy. In *Ethical and policy issues in research involving human research participants*. Bethesda, MD: National Bioethics Advisory Commission. pp. G1–G12, p. G4.
Kipnis, K. 2003. Seven vulnerabilities in the Pediatric research subject. *Theoretical Medicine and Bioethics* 24:107–120.
La Barbera, M. 2016. Interseccionalidad, un "concepto viajero": orígenes, desarrollo e implementación en la Unión Europea. *Interdisciplina* 4 (8): 105–122.
Levinas, E. 1968. *Totalité et infini: Essai sur l'extériorité*. Nijhoff.
Levine, C., et al. 2004. The limitations of 'vulnerability' as a protection for human research participants. *American Journal of Bioethics* 4 (3): 44–49.
Lombardo, E., and M. y Verloo. 2009. Stretching gender equality to other inequalities: Political intersectionality in European gender equality policies. In *The discursive politics of gender equality: Stretching, bending and policy-making*, ed. P. Meier et al., 67–83. Routledge.
Lorde, A. 1982. *Zami: A new spelling of my name*. Persephone Press.
Lorde, A. 1984 [2007]. *Sister Outsider: Essays and Speeches*. Crossing Press.
Lugones, M. 2005. Multiculturalismo radical y feminismos de mujeres de color. *RIFP* 25:61–75.
Lugones, M. 2010. Toward a decolonial feminism. *Hypatia* 25 (4): 742–759.
Luna, F. 2008. Planteos clásicos y teoría de los principios. In *Bioética: Nuevas reflexiones sobre debates clásicos*, ed. F. Luna and A. Salles. Fondo de Cultura Económica.
Luna, F. 2009a. Elucidating the concept of vulnerability. Layers not labels. *International Journal of Feminist Approaches of Bioethics* 2 (1): 121–139.
Luna, F. 2009b. La declaración de la UNESCO y la vulnerabilidad, la importancia de la metáfora de las capas. In *Análisis de la Declaración Universal de Bioética y Derechos Humanos de la UNESCO*, ed. M. Casado, 255–266. Civitas.
Luna, F. 2014. 'Vulnerability', an interesting concept for public health: The case of older persons. *Public Health Ethics* 7 (2): 180–194.
Luna, F. 2015. Rubens, corsets and taxonomies: A response to Meek Lange, Rogers and Dodds. *Bioética* 26 (6): 448–450.

Luna, F. 2019. Identifying and evaluating layers of vulnerability—A way forward. *Developing World Bioethics* 19: 86–95.
Luna, F. 2022. Vulnerability and feminist inputs. In *Routledge handbook of feminist bioethics*, ed. S. Carter et al. Routledge.
Luna, F., and S. Vanderpoel. 2013. Not the usual suspects: Addressing layers of vulnerability. *Bioethics* 27 (6): 325–332.
Mackenzie, C. 2014. The importance of relational autonomy and capabilities for an ethics of vulnerability. In *Vulnerability: New essays in ethics and feminist philosophy*, ed. C. Mackenzie et al., 33–59. Oxford University Press.
Mackenzie, C. 2016. Vulnerability, needs and moral obligation. In *Vulnerability, autonomy and applied ethics*, ed. C. Straehle, 83–100. Routledge.
Macklin, R. 2003. Bioethics, vulnerability, and protection. *Bioethics* 17 (5–6): 472–486.
Martin, A., et al. 2014. Resolving the conflict: Clarifying 'vulnerability' in health care ethics. *Kennedy Institute of Ethics Journal* 24 (1): 51–72.
Matto de Turner, C. 1889. *Aves sin nido.* Biblioteca Virtual Universal. http://www.biblioteca.org.ar/libros/71077.pdf
Mc Call, L. 2005. The complexity of intersectionality. *Signs* 30 (3): 1771–1800.
Meek Lange, R., and S. Dodds. 2013. Vulnerability in research ethics: A way forward. *Bioethics* 27 (6): 333–340.
Mohanty, C. 1988. Under western eyes: Feminist scholarship and colonial discourse. *Feminist Review* 30:61–88.
O'Neill, O. 1996. *Towards justice and virtue.* Cambridge University Press.
Ricoeur, P. 2007. Autonomy and vulnerability. In *Reflections on the just*, ed. P. Ricoeur, 72–90. University of Chicago Press.
Rogers, W. 2017. Feminist bioethics. In *The Routledge companion to feminist philosophy*, ed. A. Garry et al., 579–591. Routledge.
Rogers, W., et al. 2012. Why bioethics needs a concept of vulnerability? *International Journal of Feminist Approaches to Bioethics* 5 (2): 11–38.
Sandoval, C. 1991. *Methodology of the oppressed.* University of Minnesota Press.
Viveros Vigoya, M. 2009. La sexualización de la raza y la racialización de la sexualidad en el contexto latinoamericanoactual. *Revista Latinoamericana de Estudios de Familia* 1:63–81.
Viveros Vigoya, M. 2016. La interseccionalidad: una aproximación situada a la dominación. *Debate Feminista* 52:1–17.
Zack, N. 2005. *Inclusive feminism: A third wave theory of women's commonality.* Rowman & Littlefield Publishers.
Zion, D., et al. 2000. The declaration of Helsinki, CIOMS and the ethics of research on vulnerable populations. *Nature Medicine* 6:613–617.

Chapter 7
Vulnerability: The Practical Sphere

Abstract This chapter applies the theoretical strategy proposed in the previous chapter and shows the strength of the layers of vulnerability and their potential for harm. This chapter translates theory into practice and offers concrete guidelines for dealing with layers of vulnerability. It does not simplify or idealise research participants, research contexts or public health problems. In this sense, it accepts the challenge of providing a non-simplified response to a complex problem from the non-ideal perspective.

Keywords Vulnerability · Practical analysis · Layers of vulnerability · Stimulus conditions · Cascade vulnerability · Obligations

7.1 Introduction

This chapter develops the practical dimension of vulnerability. I focus on the second question posed in the previous chapter: the use and application of the concept.

Layered conceptual analysis has gained significant traction in both research ethics and feminism.[1] This latter recognition likely stems from my first article on the topic in the *International Journal of Feminist Approaches to Bioethics* (IJFAB), a journal edited by leading feminist bioethicists. In "Vulnerability in Research Ethics: A Way Forward" Margaret Meek Lange et al. (2013: 336) argue:

What is needed is an approach that, first of all, captures Luna's insight that vulnerable research participants inhabit a context generated by the coming together of layers of vulnerability. The approach should make some progress toward naming and classifying the layers while remaining sensitive to their possible interactions. Finally, such an account should identify vulnerability-related duties.

---

[1] See the last section "The impact of my work on vulnerability" on this chapter for details.

F. Luna, *The Ideal and Non-Ideal in Bioethics*, The International Library of Bioethics 115, https://doi.org/10.1007/978-3-032-14594-9_7

In the same vein, in "Why bioethics needs a concept of vulnerability", Wendy Rogers et al. (2012:19)[2] argue: "We also outline a taxonomy of different types of sources of vulnerability that we believe is useful to further specify the layered approach to vulnerability advocated by Luna".

Meek Lange et al. also state: "Unlike Luna's, our approach gives concrete, general guidance to researchers and research ethics committees" (Meek Lange et al. 2013: 337). These authors suggest going beyond the layers and introducing a taxonomy. Referring back to the two-sphere distinction I presented at the beginning of Chap. 6, it would seem that the conceptual sphere has acceptance and success; however, there is still work to be done in the second sphere.

Thus, while these authors appreciate my approach at the conceptual level, they want to move forward and propose an "improvement" at the practical level. It should be noted that until 2018 I had only developed the metaphor of layers; only in 2019—and as a result of these criticisms—did I formulate the concept of the stimulus or trigger condition and that of vulnerability with cascading effect presented in Chap. 6.

In this chapter I develop the practical sphere and point out how this concept should work in this area. I analyse the proposals of Meek Lange et al. (2013) but reject the model they propose as a practical way out. I question the usefulness of taxonomies for analysing vulnerabilities, and offer instead a practical solution which proposes two steps and provides a guide to help assess layers. I also identify three types of obligations and some strategies for implementing them. A feature of this practical approach is that it builds on previous theoretical developments.

7.2 Taxonomies and Corsets

The leading group of bioethicists referenced above propose a taxonomy to classify the messy layers of vulnerability. I reject this proposal (see Luna 2015) as taking us back to a rigid model.

Taxonomies[3] are introduced to understand or explain different phenomena. They originate from the biological sciences and have been used successfully in other sciences and contexts. Taxonomies may interest research ethics committees (RECs), as they provide them with a guide or checklist to analyse layers of vulnerability. However, they can also mislead, by simplifying situations that are complex. The problem lies in the difficulty of obtaining clarity and orderly classification.

Let me illustrate this with an analogy. Reality, I think, tends to resemble a woman painted by Rubens, a baroque figure who prides herself on the majesty of her voluptuous body, yet no corset—despite its supports, clamps and fabrics—will be able

[2] This team of bioethicists, in some cases with Catriona Mackenzie as well, wrote several articles along the same lines. I will focus on two articles that share the same proposition: Rogers et al. (2012); and Meek Lange et al. (2013).

[3] The authors sometimes use the term "typology", but interchange the two terms.

to contain her abundant flesh.[4] And this is what happens when we try to classify the layers of vulnerability: taxonomies, like corsets, cannot classify voluptuous reality. The real world is too complex and messy—layers of vulnerability overlap and context interacts with them. Taxonomies may work better in simpler systems, as in the ordering of some biological species, but they do not seem efficient in open and complex systems such as human behaviour and the situations to which we are exposed. Worse still, taxonomies can present the illusion of an exhaustive set of categories and suggest a false sense of order.

In my article "Rubens, corsets and taxonomies: a response to Meek Lange, Rogers and Dodds" I argued that:

a) we do not need a taxonomy to classify vulnerabilities,
b) Meek Lange et al. do not provide an adequate or accurate taxonomy, and
c) they cannot link their taxonomy to specific obligations[5] (Luna 2015).

Regarding the first two points: a) the complexity of reality exceeds the possibilities of taxonomies; and b) in Meek Lange et al.'s analysis, three categories overlap: inherent sources, situational sources and pathogenic sources. They describe these different sources as:

> Inherent sources of vulnerability include our corporeality, our neediness, our dependence on others, and our affective and social natures.... risk of harm or wrongs depends on age, health, gender and disability, as well as the person's capacities for resilience, coping and the social supports she may have. *Situational* sources of vulnerability are context specific and include 'the personal, social, political, economic or environmental situation of a person or social group'.... Pathogenic sources of vulnerability are a subtype of situational sources that arise from dysfunctional social or personal relationships. (Meek Lange et al. 2013: 336)

The problem is the overlap between inherent and situational sources of vulnerability. While inherent sources include social support or dependence on others, undoubtedly linked to the context and the particular situation of each person, situational sources are also characterised as "context specific". It is therefore not possible to distinguish clearly between one source of vulnerability and another. They end up overlapping, a fact that the authors accept. But accepting that the different sources of vulnerability overlap does not solve the problem—it only makes it explicit. If sources overlap we cannot clearly distinguish between the different categories, so the purpose of providing the order implied by a taxonomy is defeated.

I believe that overlapping is linked to their adherence to the concept of relational autonomy. This concept which has a strong presence in feminist philosophy, is incorporated within these authors' overall approach (Mackenzie 2014). But it does not seem to function adequately with the strategy of a taxonomy. I find it more compelling to maintain the concept of relational autonomy and, consequently, to abandon the proposal of a taxonomy. Finally, the third type of source, pathogenic, is a subtype of

---

[4] See Luna (2015).

[5] For a more in-depth analysis explaining the divergence between alleged sources of vulnerability and obligations, see Luna (2015).

the previous one,[6] so the proposed taxonomy ends up consisting only of two overlapping categories, meaning that it is not a truly operational taxonomy that allows us to classify and order ideas.

Regarding the third point (the linkage to specific obligations), it is not clear how each of the two categories or sources of vulnerability in the proposed taxonomy are linked to specific obligations, making this type of taxonomic proposal even more irrelevant.

However, my most important criticism is that taxonomies lead to a certain rigidity, that of establishing categories into which all vulnerabilities must fall. As we saw in Chap. 6, Kipnis (2001) proposes a taxonomy with six circumstances: cognitive, juridical, deferential, medical, allocational and infrastructural. In his taxonomy the cognitive circumstance asks whether the potential participant has the capacity to deliberate and decide whether or not to participate in the study; the juridical circumstance asks whether the potential participant is subject to the authority of others who may have an independent interest in their participation; the deferential circumstance asks whether the person-candidate has patterns of deferential behaviour that may mask an underlying unwillingness to participate; the medical circumstance assesses whether the person-candidate has been selected, in part, because they have a serious health condition for which there are no efficient remedies; the allocation circumstance asks whether the person-candidate lacks important social assets that will be provided to them as a consequence of their participation in the research; and the infrastructure circumstance asks whether the political, organisational, economic and social context of the research environment possess the integrity and resources necessary to manage the study (Kipnis 2001). The seventh circumstance which this author adds when analysing paediatric research is "social circumstances", a category that applies to all kinds of research, not only paediatric research. So, the first taxonomy proposed falls short of its aim. We can see that the addition and identification of an extra category provides another justification for avoiding taxonomies. The dynamic and flexible vision proposed by the layered approach, by contrast, shows how difficult, if not meaningless the proposed taxonomy is.

7.3 Another Practical Proposal

Although my layered approach was widely accepted by 2014, it risked paralysis for those tasked with assessing or acting on vulnerabilities. Numerous layers could be identified, but I had not put forward any criteria for prioritising and assessing vunerabilities. Taking Meek Lange et al.'s criticisms of the validity of my model into account, I formulated a practical analysis that would identify the layers and duties in a simpler and more useful way. I though of a middle way that would not include

[6] Recall that the pathogenic sources of vulnerability are those that we reviewed in the previous chapter when we analysed the layers of vulnerability with cascading effects.

taxonomies but would allow the different layers to be ordered and assessed, and might be more useful to policymakers, researchers and RECs.

There were two steps in my analysis. The first focused on identifying the different layers of vulnerability: these could be related, among others, to physical problems, consent, dependency, exploitation and socio-economic situations. Identifying the layers and their relevant features (the stimulus conditions and the cascading layers of vulnerability) would help in the assessment.

In the first step, researchers, RECs and policymakers should identify the layers of vulnerability present in a situation, research or protocol. The different definitions proposed by Deborah Zion, Samia Hurst or CIOMS (see Chap. 6) can help in this analysis. A proposal such as the one offered by Kipnis can also be useful if taken as a set of open-ended characteristics that can allow us to identify layers. However, additional features should be accepted if necessary. For example, in this first identification step we should analyse how consent is taken, whether there are social or economic situations that may trigger further vulnerabilities and thus generate other layers, whether there are gender issues that may produce further layers, and so on.

We now return to the two features of vulnerability identified in the previous section, as they can help not only to specify but also to prioritise the layers of vulnerability. In this first instance it is necessary to identify whether there are stimulus conditions or possible triggers for these layers, as well as whether some of the layers may have the cascading effect that we have seen is so damaging.

In short, this first step identifies the following:

1. The content of the layers, i.e., what the different existing layers of vulnerability consist of (we can use different definitions of vulnerability; Zion et al. 2000; CIOMS 2002/2016; Hurst 2008), as well as the characteristics proposed by Kipnis (2001, 2003).
2. Stimulus conditions, i.e., identification of triggers of the layers and the likelihood of their manifestation.
3. Cascading vulnerabilities, i.e., layers that have a cascading effect.

In the second step we need to assess how to rank and prioritise these different layers, taking into account the work from the first step. Based on the above analysis, we need to rank the layers, assessing the harms and risks involved (from physical to psychosocial, including the possibility of exploitation, dependency, abusive patterns, etc.), prioritising the most harmful layers and then moving on to the less harmful ones. As there may be several layers, we should start by assessing whether there are layers with a cascading effect. Cascading layers are often the most damaging as they have differential strength and a domino effect that can trigger multiple vulnerabilities. In prioritising cascading layers, however, we must weigh several factors—not only their harmful effect but also the likelihood of their occurrence. We need to judge the degree of harmfulness of these layers, as well as the situation in which they occur and currently existing threats.[7] This process can be understood from a coherentist model. There is no lexicographical order.

[7] For example, if we need to assess the risks that a specific protocol imposes on the research participant, we should consider whether the protocol presents minimal risk (for example, taking

The two conceptual elements introduced in Chap. 6 play a crucial role in this analysis. We need to consider the dispositional structure of the layers of vulnerability and assess which stimulus conditions can trigger them (their presence and likelihood of development). The stimulus conditions connect the layers with the context, with the actual situation and the likelihood of their occurrence. If the stimulus conditions are very likely to occur, they should be the priority target, as they are the ones that will trigger the identified vulnerability layers and cause real harm. If the conditions are related to a layer of vulnerability with cascading effect, alarm bells should ring. Although we should try to follow the hierarchy established by this process, sometimes this is not feasible, as reality may influence the possibility of actually following the established order. This hierarchy of layers should be seen as a preliminary, procedural guide for starting to design appropriate safeguards. It can by no means function as a rigid model.

Once the degree of harm and likelihood has been identified and assessed, three types of obligations must be considered.[8] Our first obligation is not to worsen the vulnerabilities of the individual or group either through a research protocol or public policy. We must avoid exacerbating layers of vulnerability. A second obligation is to eradicate layers of vulnerability; we should try to remove all layers. However, we can only require this to a reasonable and feasible extent (some layers are linked to a socio-economic context that it may not be possible to change). Thus, the third type of obligation is to minimise the layers of vulnerability. If we cannot eliminate or eradicate these layers—following the previous obligation—we should consider another alternative—finding different strategies to minimise the layers of vulnerability. We should use the ordering of the layers shown in Table 7.1 as a guide. Finally, these obligations can be expressed through different strategies, such as protections and safeguards, as well as empowerment and autonomy building.

The traditional approach to dealing with vulnerability has generally been to provide only protections and safeguards (Goodin, 1986). However, feminism has broadened this view to include the idea of empowerment, autonomy and respect for people. The importance of this is that it again avoids the idealised view of people, the assumption that some people—those who are not vulnerable—are autonomous, independent, etc., and that therefore, those who are vulnerable have certain defects and we should protect only them. On the contrary, this more dynamic vision adheres to the idea not only of protecting but also of bringing about change in people through education, raising awareness and empowerment.

I believe this process can guide the deliberative thinking and evaluation of researchers, RECs and policymakers. In this model there is no fixed priority, but a guide that should help to assess what our obligations are and where to start addressing them. It attempts to move away from taxonomy, simplification or checklists to propose a deep and serious deliberative exercise.

---

blood pressure using routine methods). In such a case, it may not be as important to eliminate the many layers compared to a research protocol that involves a liver biopsy or other risks to participants. Therefore, the contextual situation will also be relevant to our assessment.

[8] I consider and follow some of the obligations proposed by Meek Lange et al. (2013: 336–337).

Table 7.1 Step two: Assessing the practical consequences of the layers of vulnerabilities

Process of prioritisation from most to least damaging:	**Obligations:**
-Cascade effect	-Avoiding exacerbating layers of vulnerability
-Very harmful	-Eliminate layers of vulnerability
-Very likely	-Minimise layers of vulnerability
↓	↓
*** How?**	*** How?**
Identify whether stimulus conditions have been met	Through different strategies: Protections, safeguards and empowerment.

7.4 Putting the Concepts into Practice: A Case Study

Consider, for example, a drug research project involving low-income persons in countries where abortion is illegal. In this case, the possibility of pregnancy functions as a layer of vulnerability with a cascading effect. As argued in my article (Luna, 2009: 128), a woman of high socio-economic status in the same situation will also acquire a layer of vulnerability as there is a moral harm because she has to undergo an illegal abortion (even if it will probably be a safer process for her as she can pay for those safe conditions). In her case this layer will not have a cascading effect, whereas it will in the case of a low-income pregnant woman. We can easily imagine the anguish of the latter if she conceives an unwanted pregnancy in the midst of a biomedical research project testing a new drug, and her fear that the tested drug may have a teratogenic effect on the developing foetus (particularly if the intake of the new therapy occurs in the first trimester of pregnancy). If she opts for an illegal and unsafe abortion, very harmful consequences can arise: future sterility, morbidity and even mortality. But if she continues the pregnancy because she has no other choice, the baby may be born with neurological or other defects, which can lead to multiple problems. For example, she may have to provide the special care that the baby requires and will need a team that can treat the pathology adequately. Caring may take more time away from her work and income, it may generate care inadequacies for her other children as well as with her partner (it is known that when problems of this nature arise, many women are left alone in charge of the family), in addition to possible psychological problems. In other words, continuing with the pregnancy can have a series of cascading effects on the woman's life.

We must therefore identify whether there are any stimulus conditions that may actualise this cascading vulnerability and assess their likelihood. If the stimulus conditions are very likely, this layer must be prioritised. Possible actions could include guidelines to properly implement legal terminations of pregnancies or promoting campaigns to raise awareness of the issues involved, for example, for certain low-income women who lack not only socio-economic resources but also formal education and, in particular, sex education, and have no access to contraception.

Take the case of a REC that considers some research to be highly relevant to these women and their community, where other ethical conditions are also met, meriting its approval.[9] However, its members are concerned about this cascading layer of vulnerability, the likelihood that its stimulus conditions may trigger them, and how to avoid this. What are their obligations? First, they should assess whether they are exacerbating other layers of vulnerability with this protocol, then whether they can eradicate or eliminate such a stimulus condition. If they cannot, they should examine ways or strategies to minimise its occurrence. As for the possibility of eradicating it, we must recognise that without safe policies for terminating pregnancies, the possibility of having an unwanted pregnancy will always exist despite special measures to prevent it. We must also understand that researchers or RECs cannot by themselves change or enact an appropriate law or policy. (However, if this were the challenge for a public health agency, policymakers could take steps to eradicate such a stimulus condition.) Thus, inadequate sex education, lack of contraception and lack of emergency contraception are some of the stimulus conditions or triggers that can give rise to this cascading layer of vulnerability. While it may be quite difficult for RECs to eradicate this layer, they should be able to minimise it. They should consider different strategies: effective protection measures or safeguards, as well as the empowerment of these women, fostering their autonomy and respect for individuals.

For example, a first set of actions needed to achieve minimisation are: explaining and educating about methods to avoid unwanted pregnancies, educating about the use of and access to contraception (explaining that it is better to use a dual method and offering two methods of contraception, as well as the possibility of emergency contraception). They should also provide these contraception methods. They can

[9] As for example with the ethical guidelines set out by CIOMS. See CIOMS 2016 in relation to pregnant women and research (Guideline 19).

also propose other complementary actions: talking to the relevant health officials, writing articles, giving interviews to the press to raise awareness about this restrictive and harmful policy for reproductive health. These last suggestions should not be obligatory for researchers, or even for RECs, they function as supererogatory actions to be carried out to raise awareness of the problem. RECs may not be able to eradicate this layer of vulnerability, but by requiring researchers to implement at least the first set of protections mentioned (education, dual contraception and emergency contraception) they can minimise it. However, if REC considers that these strategies are not sufficient (because these women will not be able to use contraception, for example, due to refusal by their partners or because it may lead to domestic violence), the REC can either (1) reject the protocol or (2) choose not to recruit such individuals to the study and set clear conditions for the participation of others. Obviously, this last step must be carefully evaluated, as in some cases it can lead to overprotection.

7.5 Rebutting Criticism

A review of some previous critiques shows that Meek Lange et al. pointed out that my approach "should make some progress toward naming and classifying the layers while remaining sensitive to their possible interactions" (Meek Lange et al. 2013: 337). They also argued that such a proposal "should identify vulnerability-related duties" (Meek Lange et al. 2013:336). The response provided by these authors is based on the taxonomy they presented. They point out: "Unlike Luna's, our approach gives concrete, general guidance to researchers and research ethics committees" (Meek Lange et al. 2013: 337). While the authors are right to criticise the lack of concrete guidance in my initial proposal, as I have argued, I do not believe that classifying the layers in a taxonomy is the relevant task at hand. Moreover, the difficulty of distinguishing between the two categories presented in the proposed typology renders this taxonomy unsuccessful. The identification of taxonomies or types of vulnerabilities is not the priority, particularly if it is not clearly linked to specific obligations that really contribute to a practical response.

On the contrary, especially for operational purposes, what matters is the normative force of harmful layers and the avoidance of damage. In addition to my theoretical proposal defending the relevance of using layers of vulnerability and thinking of them in a dynamic way, as far as a practical response is concerned, I highlight the two relevant theoretical features presented, which have practical impact: their dispositional character and the relevance of stimulus conditions or triggers, and the possibility of them having a cascading effect. My proposed two steps—identification, and evaluation and prioritisation—are used in the assessment process and I recognise three types of obligations. My proposal is based on analysing the severity and likelihood of vulnerabilities to be triggered, given that they imply harm, risks or threats to individuals or groups. The relevance of considering cascading vulnerability lies in being able to anticipate the possible generation of new layers and risks. For example, in the case of public health ethics and research, RECs, researchers or legislators, rather than being concerned with identifying a category in a taxonomy (especially when such a typology is not clear, nor leads to a direct application of obligations), they should seek to avoid, eradicate and minimise layers of vulnerability.[10]

While I offer guidance, it is still essential that a thorough examination be undertaken. My proposal forces moral deliberation and avoids simple and automatic analysis through the use of checklists. But, at the same time, it provides valuable guidance on how to prioritise criteria and what to try to address first. I believe that RECs, like those who legislate or administer in the case of public health, must conduct a thorough ethical analysis. They cannot rely on a list of "supposedly" vulnerable populations because, as has been shown, this risks stigmatising these groups.

My current strategy offers a workable and practically useful response by suggesting a middle way between a rigid taxonomy and the many messy layers. It integrates standard definitions and accounts of what is considered to be vulnerability (Zion et al. 2000; Hurst 2008; CIOMS 2002; Kipnis 2001, 2003). It highlights the relevant features of the layers as provisions and provides a ranking and assessment process with explicit obligations to guide RECs and those who research, legislate or administer. This analysis generates obligations, but is clearer and simpler than the taxonomy-based approach. It provides concrete guidance. It brings substantive content to the practical domain, but does not simplify or idealise research participants, the research context or public health challenges.

7.6 The Impact of my Work on Vulnerability

My conception of vulnerability has been significantly influenced by feminist bioethics, with which I established a fruitful dialogue and highly engaging discussions. Respected bioethicists and philosophers such as Catriona Mackenzie, Susan Dodds, and Wendy Rogers were early adopters of the metaphor of layers and went beyond it with their own proposals. Like me, they worked on the development and conceptualisation of this concept. Although I do not fully agree with some of their theoretical developments, many others influenced me or proved highly valuable. My concept of cascading vulnerabilities was based on their formulation of the subcategory of pathogenic source (although I diverge from their analysis, I must acknowledge that their approach helped me discover the rich notion of layered vulnerability with cascading effects). Thanks to their critiques and friendly comments, I have continued exploring its normative power and practical implications, seeking a practical response to the metaphor of layers while developing the dispositional and trigger-based approach that plays a central role in evaluating those layers. There is, then, a genuine dialogue and mutual enrichment between our frameworks. For It was an honour to be invited to contribute the chapter on vulnerability (Luna 2022) to the paradigm-shifting *Routledge Handbook of Feminist Bioethics*, co-edited, among others, by Wendy Rogers (Carter et al. 2022).

[10] See my criticism in Luna (2015: 450).

My proposals have also influenced many other feminist bioethicists who have applied them to their areas of interest, including Elizabeth Victor, Alison Reiheld and Laura Guidry-Grimes. With them, I co-presented a panel at the International Association of Bioethics (IAB) 2020, and we later published an article applying not only the notion of layers but also several concepts from my proposal to various practical domains (Victor et al. 2022).

Another area where my proposal had significant impact was the development of research ethics guidelines At the 2016 CIOMS presentation, the President and the Scientific Secretary explained why there was a need to revise the 2002 Guidelines (Van Delden and Van der Graaf 2017). Among other reasons, they recognised the need to clarify how the concept of vulnerability should be understood. The 2016 version of the CIOMS guidelines retains the 2002 definition, but includes my proposed layers. Although it does not use the word 'layer', It follows my ideas and moves away from the traditional sub-population approach.

The commentary to CIOMS Guideline 15 states:

> A traditional approach to vulnerability in research has been to label entire classes of individuals as vulnerable. The account of vulnerability in this guideline seeks to *avoid considering members of entire classes of individuals as vulnerable.* However, it is useful to look at the *specific characteristics* that may render individuals vulnerable … *Different characteristics may also co-exist*, making some individuals more vulnerable than others. This is *highly dependent on the context.* For example, persons who are illiterate, marginalised by virtue of their social status or behaviour, or living in an authoritarian environment, may have multiple factors that make them vulnerable. CIOMS (2016: 57)

Several of the elements addressed in this book can be identified in this paragraph: the idea of avoiding labelling; the need to refrain from categorising entire classes of individuals (subpopulations) as vulnerable; and the coexistence of different characteristics and multiple factors that can render individuals vulnerable, which clearly aligns with the metaphor of layers. Finally, it also emphasises how strongly context dependent these factors are. Undoubtedly, the metaphor of layers is present in these important research ethics guidelines.

A similar change occurs in the 2024 revision of the Declaration of Helsinki (DoH, WMA 2024).[11] Paragraph 19 states: "Some individuals, groups or communities *are in a situation of more vulnerability* as research participants *due to factors that may be* fixed or *contextual and dynamic*, and thus are at greater risk of being wrong or incurring harm" (DoH, WMA 2024).

Both these research ethics documents are consistent, insofar as they avoid essentialist thinking and use the same conceptual scheme. Note that in the DoH 2024 no subpopulations or vulnerable groups as such are mentioned, but "situations of vulnerability" are—situations that can occur due to factors that may be fixed or contextual and dynamic.

Thus, in a Guest Editorial published in *Bioethics,* I argue there is a paradigm shift into the inclusion of this thinking:

[11] In May 2024, I was invited—alongside Samia Hurst and Alex London—to an internal meeting in Munich to discuss the notion of vulnerability in the new version of the Declaration of Helsinki.

But the 2024 DoH expands and refers to "a situation of more vulnerability as research participants, due to factors that may be fixed or contextual and dynamic", thus introducing a different way of thinking about the concept that captures the various ways in which vulnerabilities are expressed. (Luna 2025).

This new conceptualisation provides a better tool to analyse the real world and the complexities it entails. In addition, it fosters deliberation and critical thinking.

7.7 Conclusion

This chapter complements the previous chapter by showing how the theoretical elements analysed at the conceptual level are relevant when thinking about practical application. This underscores the importance of rigorous theory in bioethics—it is not merely abstract philosophising. This chapter reviews the criticisms raised by some respected colleagues. It takes their judgements seriously and shows why the alternative they propose does not work. In contrast to the taxonomic model, practical work is proposed to identify and assess the different layers, taking into account theoretical concepts such as the dispositional nature of vulnerability, meaning that there are stimulus conditions that can actualise vulnerability, or that some layers of vulnerability can have a cascading effect and be terribly damaging. A case study clarifies the proposal. Finally, the initial criticisms are analysed to show that my current proposal is superior not only to the approaches of Meek Lange, Rogers and Dodds, but also to my own initial approaches which only identified multiple and messy layers, and did not provide more precise guidelines for their assessment.

References

Carter, S., et al., eds. 2022. *Routledge handbook of feminist bioethics.* Routledge.

Council for International Organizations of Medical Sciences (CIOMS) in collaboration with the World Health Organization (WHO). 2002 (2016). *International Ethical Guidelines for Health-Related Research Involving Human Subjects* (3rd ed.) Geneva: CIOMS.

Goodin, R. 1986. *Protecting the vulnerable: A re-analysis of our social responsibilities.* University of Chicago Press.

Hurst, S. 2008. Vulnerability in research and health care; describing the elephant in the room? *Bioethics* 22 (4): 191–202.

Kipnis, K. 2001. Vulnerability in research subjects: A bioethical taxonomy. In *Ethical and policy issues in research involving human research participants.* Bethesda, MD: National Bioethics Advisory Commission. pp. G1-G12, p. G4.

Kipnis, K. 2003. Seven vulnerabilities in the Pediatric research subject. *Theoretical Medicine and Bioethics* 24:107–120.

Luna, F. 2009. Elucidating the concept of vulnerability. Layers not labels. *International Journal of Feminist Approaches of Bioethics* 2 (1):

Luna, F. 2015. Rubens, corsets and taxonomies: A response to Meek Lange, Rogers and Dodds. *Bioética* 26 (6): 448–450.

Luna, F. 2022. Vulnerability and feminist inputs. In *Routledge handbook of feminist bioethics*, ed. S. Carter et al. Routledge.

Luna, F. 2025. Editorial Guest: Vulnerability: on a welcome paradigm shift in the 2024 Declaration of Helsinki. *Bioethics*.

Mackenzie, C. 2014. The importance of relational autonomy and capabilities for an ethics of vulnerability. In *Vulnerability: New essays in ethics and feminist philosophy*, ed. C. Mackenzie, W. Rogers, and S. Dodds, 33–59. Oxford University Press.

Meek Lange, M., et al. 2013. Vulnerability in research ethics: A way forward. *Bioethics* 27 (6): 333–340.

Rogers, W., et al. 2012. Why bioethics needs a concept of vulnerability? *International Journal of Feminist Approaches to Bioethics* 5 (2): 11–38.

Van Delden, J., and R. Van der Graaf. 2017. Revised CIOMS international ethical guidelines for health-related research involving humans. *JAMA* 317 (2): 135–136.

Victor, E., F. Luna, A. Reiheld, and L. Guidry-Grimes. 2022. Vulnerability in practice: Peeling back the layers, avoiding triggers, and cascading effects. *Bioethics*, Special issue: IAB 15TH World Congress 36 (5): 587–596.

World Medical Association (WMA). 2024. *Declaration of Helsinki: ethical principles for medical research involving human subjects*. Adopted by the 18 WMA General Assembly, Helsinki, Finland, June 1964, 8th ed., 75 WMA General Assembly, Helsinki, Finland, October 2024.

Zion, D., et al. 2000. The declaration of Helsinki, CIOMS and the ethics of research on vulnerable populations. *Nature Medicine* 6:613–617.

Chapter 8
Relational Autonomy: Another Non-Ideal Analysis

Abstract The last chapter deals with another fundamental concept for bioethics, presented in a non-ideal key: relational autonomy. The chapter shows how feminism has highlighted the implications, commitments and applications of this concept with an emphasis on how our non-ideal world leads to failures of individuals and groups to exercise their autonomy. This chapter not only shows the different versions of hegemonic feminism, but also presents the vision of feminism from the South. Another important issue this chapter outlines is that faced with the opposition that is traditionally posed between vulnerability and autonomy, it proposes to rethink them from more sophisticated perspectives that avoid this dead end. The chapter shows how relational autonomy can complement or be constructively related to the notion of vulnerability and provide more complete and interesting analyses for medical practice.

Keywords Relational autonomy · Feminism · Non-hegemonic views · Relational autonomy and vulnerability

8.1 Introduction

This chapter is devoted to the concept of autonomy, a fundamental principle of ethics that has been the subject of substantial developments in feminism.

First, autonomy is far more complex and problematic than typically assumed. When examined through both ideal and non-ideal frameworks, its analytical depth becomes evident. Feminism[1] has highlighted the implications, commitments and applications of this concept with an emphasis on how our non-ideal world leads to failures of autonomy—although feminists are not its only critics. Feminism accurately points out bad idealisations such as the idealised view of human agency. Importantly, these shortcomings or problems are present in traditional analyses, especially within

[1] Although feminism is referred to here in the singular, there are of course multiple versions of feminism (liberal feminism, egalitarianism, radical feminism, etc.), and within each there is also a variation of proposals.

F. Luna, *The Ideal and Non-Ideal in Bioethics*, The International Library of Bioethics 115, https://doi.org/10.1007/978-3-032-14594-9_8

bioethics, which tend to analyse the concept of autonomy according to idealised visions. Autonomy thus turns out to be a battleground of the ideal and non-ideal spheres.

A second point is that the concept of autonomy often operates as a counterpoint to vulnerability. According to some traditional interpretations, if a person is not autonomous, they are vulnerable (Fineman 2010; Álvarez Medina 2022). This sort of antagonism is wielded, above all, from traditional views. For example, Levine et al. (2004) assume this notion of autonomy and end up this article with a restrictive conception of vulnerability (one that arises only when there is no autonomy). These conceptualisations propose a counterpoint relationship between the two concepts (Álvarez Medina 2022): the need for protection is posited precisely because there is a lack of autonomy and the person is labelled as vulnerable. When vulnerability is associated with the loss of certain intellectual capacities or related problems, the traditional 'obstacles to autonomy' (non-rationality, mental pathology, ignorance) are posited as revealing vulnerabilities.

This chapter reviews some of the criticisms of this concept. The bad idealisations identified by some feminist scholars can clearly be linked to the non-ideal, given feminism's sensitivity to context and oppression. The following sections present different versions and demands formulated by relational autohony as well as some formulations more closely linked to a Global South perspective. We will revise how Martha Fineman's position emphasises the notion of vulnerability over that of autonomy. Catriona Mackenzie, however, rescues the concept, showing the importance of the principle of autonomy for feminism. The chapter ends with a return to concrete practice, offering a complementary rather than antagonistic reflection on the concepts of vulnerability and relational autonomy.

8.2 Feminist Critique

From the 1990s onwards, different versions of feminism began to revise and question the concept of autonomy. Feminist critiques highlight how traditional personal autonomy relies on methodological individualism, isolating individuals from their social and historical contexts (Christman 2004). This reflects an idealised view of human agency. Another point feminists make is that traditional formulations overemphasise freedom. This can be traced back to Kantian ethics or the Rawlsian formulation that is based on free and rational agents (Stoljar 2013). An average, seamless citizen is assumed and, as Álvarez Medina states, this model provides the illusion of full autonomy (Álvarez Medina 2022). In this traditional conception we can again see an idealised reading of autonomy, a conception of a person that does not exist, since we are not isolated, entirely independent, or merely rational agents. This way of considering autonomy has aspirational elements. It is a non-existent ideal and, even if many philosophers and defenders of this model recognise that autonomy occurs in degrees and that there is no such thing as 'complete' autonomy, this ideal continues to operate as a basic standard in both bioethics. Mainstream bioethics has not been enriched by inside ethical discussions and an over simplified analysis of authonomy has been incorporated. This simplistic conception affects the practical application of the concept, because people's autonomy does not seem to operate in isolation and independently of others, but is exercised through and influenced by social relations and ties (Nedelsky 1993; Butler 2009; Buedo and Luna 2021). Most people, for example, faced with a decision whether to undergo surgery, consult and consider their loved ones, who can take care of any dependent family members, implications for their work situation, etc. These criticisms bring us back to Robeyns' analysis of good and bad idealisations and to the need to filter out bad idealisations.[2]

Another element playing into this conception of autonomy that is significant for feminism is the split between reason and emotion, between the individual and the relational, which is linked to the distinction between the public and the private (Álvarez Medina 2022: 13). Such differentiations, in illustrating a way of ordering society, indicate a certain patriarchal spirit, appearing to support the typical masculine ideal centred on the rationality of agency. Autonomy is presented as that which defines the person, their motor, their primary way of interacting with other people and with their environment (Álvarez Medina 2022: 12).

Feminist ethics questions the limitations of this traditional liberal approach to the concept of personal autonomy and its assumptions (Álvarez Medina 2022; Ramón Michel and Cavallo 2022; Belli and Suárez Tomé 2021). Interestingly, feminism considers other examples of failures of autonomy linked to practices of gender oppression that are not sufficiently highlighted in traditional bioethical analyses: people who are very self-sacrificing or excessively deferential to others; or practices or preferences that are adapted to accommodate oppressive social conditions or decisions that accept questionable practices such as the use of the veil or female genital mutilation (Stoljar 2013). Thus, feminism brings oppression, context, society, and social bonds into the debate, and in doing so, opens up the discussion on the concept of autonomy. It also brings in a non-ideal perspective.

8.3 Reclaiming the Concept of Autonomy

A leading philosopher in the development and analysis of this concept is Catriona Mackenzie (2014, 2018, 2022). She argues for the importance of maintaining autonomy in the face of the onslaught of another renowned feminist philosopher, Martha Fineman. Both authors analyse the concepts of autonomy and vulnerability and dive into their relationships and implications. Both interpretations hold an ontological view of vulnerability. But Fineman goes further and defends the model of a vulnerable subject. She argues that vulnerability is universal (an inherent human condition) yet manifests in particular ways: in other words, we are all vulnerable, bodily and constantly, but each individual is vulnerable in a different way. It is an

[2] See Chap. 2.

"inevitable, enduring aspect of the human condition" (Fineman 2008: 8), an ontological condition of our embodied humanity. Fineman emphasises that vulnerability arises interactively, shaped by environments that render individuals subordinate or dependent. For her, if the social aspect or the structural matrix of vulnerability is not taken into account, there is a risk of focusing on individual deficiencies and forgetting the processes and interactions they originate from, such as the systems of oppression, inequality or domination that are present in different societies (Álvarez Medina 2022: 16).

Fineman thus posits another way of reading idealisations. For her, the libertarian vision of autonomy characterised as non-interference is a response to what she calls "the myth of autonomy" (Fineman 2008, 2010). This myth is central to the libertarian model of the subject-state relationship, based on a skewed conception of citizenship as self-interested, independent and rationally contracting. The problem Fineman points out is that this threatens the development of legal, institutional and social structures that should rectify disadvantage and inequality. Thus, she reconceptualises the responsibility of the state and what we are owed as citizens, proposing the vulnerable dependent subject as in need of protection, and allowing us to think differently about the role of the state as both provident and protective. In this reading, the duties of the state would derive from the principle of equality, while in the traditional autonomy-based model, the state is expected to guarantee the absence of interference in the exercise of autonomy (Fineman 2010). This approach, moreover, does not stop at the individual approach to autonomy but shows how it impacts on our conceptualisation of society and the duties of its institutional structures.

Catriona Mackenzie agrees with Fineman's critique of libertarianism and with the ontological view of vulnerability.[3] A number of feminist thinkers incorporate an ontological view of vulnerability linked to the body and finitude, an essentialist interpretation with a flavour of the continental philosophy that naturalises vulnerability. Although Fineman does propose a more egalitarian view and offers a valuable attempt to show another possible society, she remains trapped in a traditional interpretation of autonomy. As we shall see, autonomy does not necessarily have to be exhausted in the individualistic view that Fineman espouses. Mackenzie (2022: 54) also criticises the methodological individualism that underlies traditional views of autonomy and points to its idealised character, but—unlike Fineman—defends the normative importance of autonomy.

In contrast to Fineman, Mackenzie (2014, 2018) defends the concept of autonomy as allowing us to assert ourselves as decision-makers. Note that the concept of autonomy provides a certain power that means it cannot be dismissed so quickly. For Mackenzie, this is a crucial component of leading a fulfilled life. Mackenzie points out that some level of self-determination, self-governance and self-empowerment is

[3] As we have seen, in my opinion, it is precisely this adherence to the ontological conception of vulnerability that will create problems for her when defending a taxonomy of vulnerabilities, as it is almost impossible to delimit what corresponds to the inherent and what to the situational source of vulnerability (see Chap. 7 and Luna 2015).

crucial to women's emancipation, and that gender oppression restricts women's abilities to lead self-determined lives, impairing their capacities for self-governance and undermining their self-perception as self-empowered agents (Mackenzie 2022: 50–51). And logically, she concludes by stating that a reworked conception of autonomy is needed. Non-ideal lenses show these potential oppressions. They ought to be taken seriously into account, alongside an analysis of the playing out of society and context.

These considerations are in line with Natalie Stoljar's (2000) proposal. Stoljar stresses the importance of a substantive theory that is compatible with what she calls a feminist intuition. Such an intuition implies that moral agents can see themselves as sources of autonomy, that they can respond in ways that are akin to their self-confidence, self-worth and self-esteem, and that they feel subjectively able to account for their action or choice and can be held accountable for it.

A shift in the focus of the discussion that needs to be taken into account is that many approaches move away from the analysis of autonomous actions or autonomous decisions. Such approach is commonly in bioethics and in the medical field, whether clinical or research (e.g., when examining the informed consent process). Indeed, Stoljar (2000) speak of a decisional autonomy linked to informed consent and take Tom Beauchamp (Beauchamp and Childress 1999) as their representative. In these new conceptualisations, on the other hand, the autonomous person is considered. This latter position is more akin to law, since it considers the person in general and their capacities (testamentary capacity, etc.). It is also in line with feminist advocacy that seeks to reverse the situation of women in society. Such a rethinking of the approach is relevant because the shift from 'actions' to 'persons' changes the conditions and demands as well as what such concepts imply.

8.4 Different Proposals

We have noted the importance of maintaining autonomy in our lives, but new models are required. Thus, a valuable and interesting concept from the non-ideal arena proposed by feminists is the notion of relational autonomy. The various different conceptions of relational autonomy share one overriding characteristic: that social relationships are central to the exercise of autonomy. Relational autonomy allows us to think about which social circumstances enable a person to exercise autonomous actions, or which would limit them (Butler 2009). It considers the context of decision making, substituting a more realistic view for the purely ideal and fictional model.

One of the first feminist philosophers in the world of bioethics to begin working with this concept organically is Susan Sherwin (1998, 2014). She enriches the notion of autonomy and avoids simplifying social relations, pointing out that such relationships can be a double-edged sword within a patriarchal culture. But she also sees them as a tool to combat certain attitudes such as medical paternalism.[4]

---

[4] At the 1994 International Association of Bioethics Congress in Buenos Aires, we served on the same panel. Sherwin questioned the viability of the concept I presented in a paper (Luna 1995) on

This concept can also be linked to a controversial theoretical topic within feminism: the ethics of care.[5] For if one follows Stoljar's analysis, one can establish certain relations with this theoretical proposal. Stoljar analysis, set out in the *Stanford Encyclopedia of Philosophy*, is relevant:

> The term "relational" here may serve simply to deny that autonomy requires self-sufficiency. If relationships of care and interdependence are valuable and morally significant (cf. Mackenzie & Stoljar 2020, 8–10), then any theory of autonomy must be "relational" in the sense that it must acknowledge that autonomy is compatible with the agent standing in and valuing significant family and other social relationships. (Stoljar 2013).

However, Stoljar dismisses this meaning. In most feminist proposals, the emphasis will be on the relational in that we are socially and historically situated persons and this has implications within the moral sphere and in decision making.

Mackenzie develops this concept and its implications in depth. She understands autonomy as the capacity to lead a self-determined life as well as being recognised as an autonomous agent by others. For Mackenzie, relational autonomy is a multidimensional, not a unitary concept. It involves three interrelated dimensions:

(1) self-determination as freedom of opportunity. It identifies external and structural conditions for individual autonomy in which freedoms and opportunities or the lack of them are at stake;
(2) self-governance linked to the agent's internal competence, capacity and authenticity of decisions (a socially constituted capacity);[6] and
(3) self-realisation as the self-recognition of decision making and being able to take responsibility for it (Mackenzie 2018).

This position implies possessing some attitudes of self-assessment, self-respect and self-confidence. It should be noted that the last two dimensions are conditioned by the socio-relational variable, and that the emphasis is on the autonomy of the agent.

In Mackenzie's proposal, relational autonomy is not intended, as a theoretical and practical concept, to be somehow opposed to personal autonomy. This philosopher is committed to normative individualism in which rights, welfare and autonomy impose normative constraints on collective claims (Mackenzie 2022: 51). The approach seeks to ensure the existence of personal autonomy, and to add socio-relational conditions as a fundamental variable for its implementation. Of course, this is a gradual concept: a person can be self-determined, self-governed and self-fulfilling to different degrees

---

respect for illiterate women and their autonomy in the clinical setting (I pointed out the fallacy of denying them autonomy because of their inability to read or write, and the importance of being able to respect them as moral agents). Years later, she told me that approaches like mine made her desist from discarding this important concept.

[5] Some views label this position as essentialist. They highlight aspects that are the result of women's subordination and call it "feminine" ethics (Nelson 1992); others stress the importance of responsibility towards other people and rescue it as valuable (Kohen 2005). Gilligan herself, a founding theorist, tries to avoid an essentialist interpretation (Gilligan 1985: 14). For more on these approaches see Salles (2008: 113–126).

[6] This is where interesting debates arise regarding the possibility of autonomy in oppressive contexts as well as adaptive preferences (Mackenzie 2022; Cavallo and Ramón Michel 2022).

and this can change throughout life. And because it is not unitary, it allows for a certain flexibility according to the individual.

Mackenzie provides examples from public health to show how policies that for traditional autonomy may not be considered paternalistic, for a relational autonomy view they do. And she also shows how this is an innovative approach to traditional philosophy (Mackenzie 2018). For example, she assesses the Australian LiveLighter television advertisement campaign showing people's internal organs surrounded by a kind of unpleasant yellow liquid representing toxic fat. Such a campaign suggests that remembering this repulsive image will help people control their unhealthy eating habits if they are tempted to eat pizza, chips or a sugary drink. Carter et al. (2015), applying Mackenzie's proposal, point out that this campaign, instead of improving autonomy, is boycotting it. These authors argue that from a liberal perspective,[7] this type of advertising respects autonomy as it simply provides information about health, enabling people to make an autonomous decision. However, they show how this campaign reinforces the social stigmatisation of obesity by implying that large or bulky bodies are disgusting and that what makes people overweight is their lack of self-control. This type of message undermines self-governance and self-empowerment as it is likely to generate more anxiety, lower activity and motivation, and damage self-esteem and self-confidence. The authors point out that such campaigns need to ensure the messages they are sending do not boycott self-empowerment.

Along the same lines, Diana Meyers (2000) points out, in addition to self-confidence for self-definition, the need for certain skills. She proposes considering introspection skills to interpret one's own subjective experience, and imagination and memory skills to retrieve important experiences. It includes communication skills to exchange insights, knowledge or advice with others as well as analytical reasoning skills to be able to compare and assess different self-concepts and alternatives. The proposal also takes into account willingness to resist pressure from others and the interpersonal skills to make alliances that challenge and change social norms (Meyers 2000). Mackenzie also adheres to this line and considers self-definition as a fundamental capacity. Note how these approaches include a range of situations, competences and capacities that must be present when talking about autonomous actions and people.

---

[7] Remember that there is a whole line of liberal feminist philosophy (Stoljar and Mackenzie 2022).

8.5 Levels of Requirements

When examining the different relational proposals, it can be noted that there are different levels of demand. In this sense, another relevant distinction here is that between autonomy defined in procedural or substantive terms (Stoljar 2013).[8] Procedural approaches limit the necessary requirements to competencies or intellectual capacities for reflecting on one's motivations, beliefs and values—focusing primarily on specific steps or procedures that must be followed.[9] Substantive approaches impose normative constraints on autonomy. They imply values.

Thus, procedural proposals are content neutral: they do not define a type of preference or a certain type of decision, but assess whether a certain procedure took place in order for the decision to be regarded as autonomous (Ramón Michel and Cavallo 2022: 32).

For some feminists within this line of thinking, some decisions could be genuinely autonomous if certain requirements are met, including subjecting their beliefs to some critical adherence (Ramón Michel and Cavallo 2022: 32). They point out that people's life plans and conceptions of the good life are influenced by a variety of social factors such as class, gender, and so on, arguing that theories of autonomy should be neutral in relation to life plans or conceptions of the good. According to some feminists, a theory of autonomy should not homogenise people (Meyers 2000). Nor should feminist, liberal or any other ideals be imposed in the name of autonomy (Stoljar 2013).

Among the representatives of this relational procedural position are Marilyn Friedman (2003) and Diana Meyers (2000). For them, autonomy is achieved when the person undergoes a historical process of critical reflection. For example, Friedman speaks of a reflexive endorsement, a strong commitment. She refuses to treat people in an abusive domestic relationship due to adherence to oppressive religious or moral norms as autonomous in the highest degree, proposing instead a threshold that must be met. In this case, Friedman argues that people are locally autonomous in relation to these preferences (Friedman 2003). She points out that autonomy may not be the only value and that such cases can be explained through a range of values, some of which are of great importance to the individual. Friedman regards future autonomy as subordinate to other considerations. She considers people as less autonomous insofar as they adopt norms that devalue autonomy (Friedman 2003) but continue to make a locally autonomous decision.

While these procedural relational theories are interesting and have many strengths, they have been criticised for not placing adequate weight on the effects of internalised oppressions on people's motivational states, and for not sufficiently taking into account the constitutive role of external conditions (Stoljar 2013). Note that oppressive choices such as those just described call into question the very essence of feminism and its critiques of patriarchal societies.

---

[8] Stoljar also distinguishes between causal and constitutive relational theories. I will not present them here. See Stoljar (2013).

[9] In the 1970s some of its representatives were Gerald Dworkin (1988) and Harry Frankfurt (1988).

Within the substantive positions there are weak and strong positions. The weak ones demand normative content, but not directly on the subject's preferences or values, but on some competences such as self-respect in the style of Andrea Westlund (Westlund 2022; Ramón Michel and Cavallo 2022). For strong substantive positions, on the other hand, the content of the preference or values that the person forms are subject to normative constraints (Benson 2005 in Stoljar 2013). In this analysis, the preference to be enslaved or subservient cannot stand alone. Certain preferences or values are incompatible with autonomy, not because of how they are formed but because of their content.

There are different degrees of strong substantivisms.[10] Strong substantivists like Marina Oshana require certain content in people's own values and preferences, so that one will be autonomous if one critically adheres to moral commitments and preferences that are not incompatible with an autonomous life (Ramón Michel and Cavallo 2022: 32). People living under conditions of severe social limitation cannot be autonomous. For Oshana, autonomy should be among the things that are valued, or values should be in line with it (Oshana 2022). But, furthermore, for this philosopher autonomy requires certain conditions of social justice, in addition to people having practical control over their lives. It demands that the subject values her autonomy. The self-sacrificing, deferential or Taliban woman is not autonomous. She describes the latter as follows:

> She is not allowed to support herself financially. She does not have legal custody of her offspring—who are left in the hands of their father or male relatives if he dies. She has no say in the form and duration of her children's (particularly her female children's) schooling. She must wear uncomfortable clothing (the burqa) when in public. She may not travel unless accompanied by a male relative… She knows that any transgression and show of independence counts as heretical defiance and invites light or harsh punishment. (Oshana 2022: 60).

For Oshana, therefore, a woman who does not have de facto control over routine aspects of her daily life cannot be autonomous.

This strong position has also been the subject of numerous criticisms. Some objectors point out that it contradicts the aspiration to respect the different life plans of each person; or that it may overlook the efforts that people make to exercise autonomy, however imperfectly; or that it simply ends up distorting the very notion of autonomy itself (Oshana 2022). Note that these debates are not exhausted by the basic proposition of mere decision making in the abstract, but have as their backdrop the various forms of oppression in which some women find themselves and how autonomous they may be. And this can range from the model of the self-sacrificing woman who subordinates her will to the will and desires of others to those who live in authoritarian regimes that leave them very little space for decision making.

The applicability of this concept is also defended from substantive positions (Oshana 2022). And while it is a very interesting concept, it also suggests a rather demanding vision that can be very perfectionist and less inclusive than desired. For all these reasons, I agree with Natalie Stoljar that there is no consensus as to which

[10] See Oshana note 2 for the range of relational approaches (Oshana 2022:125).

theoretical position is correct. There is, however, some agreement that oppressive socialisation practices and oppression itself diminish autonomy and perhaps even eliminate it (Stoljar 2013).

8.6 Some Proposals from the South

At this point it is worth rescuing the view of these questions from the South, where there is a large and diverse literature to draw upon.[11] Among the valuable approaches from Latin America, I will present some ideas of just one of its representatives (Lagarde 1997). Southern feminist thought challenges hegemonic critiques and redistributes epistemic authority. There are thus strong similarities with the approaches we pointed out regarding intersectionality and the 'alienated' gaze from the periphery of black feminism. Latin American women are also the 'others' of hegemonic feminism: Indigenous and peasant women, Afro-descendants, mestizas, migrants, lesbians, on the periphery, in post-colonial societies, etc. (Bidaseca and Vazquez Laba 2011). While there is a clear critical stance from some Southern feminisms on the ethnocentrism and limitations of modern Western liberal feminism, there is a range of heterogeneous feminisms (autonomous institutionalist, indigenist, Afro, lesbian, decolonial, etc.). Each of them addresses key questions in a different and specific way. Thus the feminism that derives from the genealogy of Southern women implies a critical reflection on the questioning already formulated by 'mainstream' feminism (Ciriza 2015; Espinosa Miñoso 2017; Alvarado 2020).

One conceptualisation of autonomy that I would like to highlight as another alternative to the prevailing feminist views in the North plays with the notion of bodily autonomy.[12] This is related to feminist activism in relation to sexual and reproductive rights (which we have already seen is a pending issue in almost the entire region), to the militancy for diversity, to 'fat' activism, among others. These question the pathologisation of bodies and the neoliberal ideological position that is based on stereotypes of female bodies. Thus, what is postulated as healthy is questioned and it is pointed out that the right to decide for oneself is expropriated (Morelli 2022: 70; FSS 2019). This proposal advocates a redistribution of epistemic authority that is very interesting in its practical application. For this type of proposal, epistemic authority must return to those who possess such bodies. For example, in the medical field, a horizontal dialogue with the medical institution is proposed. An analysis that can complement or continue this is that of Valls-Llobet (2008), who analysed four types of violence that tend to occur in women's health care contexts. This type of approach is relevant in cases where women are reduced to their biological bodies and

[11] For example, there are analyses much more closely linked to the political realities of Southern societies characterised, among other things, by inequality, colonisation and violence. All these points are beyond the scope of this chapter.

[12] I take much of this analysis from the excellent thesis for the Master's degree in bioethics at FLACSO presented by Sabrina Soledad Morelli in 2022 (Morelli 2022, unpublished).

lack authority, a situation that is relatively frequent in health care settings, especially in the case of women on the periphery.

Lagarde (1997) is a Latin American feminist thinker who works on embodiment and autonomy. For her, autonomy is constructed from different vital processes and includes the sexual process. She points out that in the case of women it is a double sexual process because women's definition of gender is based on the mutilation of their sexual autonomy. Thus, constructing autonomy implies critically reviewing one's own sexuality in order to transform its content. Lagarde starts from the basis of a lived body making the construction of our autonomy possible. She formulates a critique of the ethics of care: the neglect of self-care. Thus, women's self-care should be at the centre of the construction of autonomy that she claims. For this author, the realisation of autonomy in women is related to biography as a method: to the possibility of critically thinking about one's own life. There would be no possibility of autonomy when one's own life cannot be reflected upon, when there can be no biographical analysis and appropriation. To avoid the criticism that her approach is reductionist, she argues that "the subjectivised body is affective, symbolic, thought, etc. It is also culturally constructed. Moreover, it is culturally constructed and psychological processes are also cultural, social, economic and political" (Morelli 2022). Finally, and as Morelli (2022) points out, Lagarde clarifies that "approaching our experience of the 'lived body' critically and biographically would imply recovering our body which, in principle, is constituted as a body for others: the procreative body for the life of others, the nourishing body for the life of others, the erotic body for the pleasure of others, the aesthetic body for the enjoyment of others, etc." (Lagarde 1997). These approaches also incorporate the social and, in a way, relational aspects, since this author puts the body in a continuous critical dialogue in its link with 'other people'. She underlines the impact that these people have on each one of us. Note the strength and originality of Lagarde's writings as they were published in the 1990s. Her work also has a fundamental relevance to women's struggle for respect for their reproductive rights, including the view of one's own body and the possibility of appropriating and empowering oneself.

The new conceptualisations of the concept of autonomy are thus extremely varied and rich. They range from clearly relational approaches, analyses that imply a series of demands or requirements that are not only cognitive but also volitional, to proposals that focus on the body and its connection with society through the body. There are approaches from mainstream feminism and from the periphery (from those that are not so accepted and that do not consist of what society considers "normal"). However, they undoubtedly exhibit the richness and variety that we find in the non-ideal arena.

8.7 Vulnerability and Autonomy in Mentally Ill Patients

At this point it is worth looking at whether there are spaces in which the concept of relational autonomy can be linked to that of vulnerability, and the impact of reconceptualising 'vulnerability' and 'autonomy' through a non-ideal lens. I argue that instead of subscribing to a counterpoint and opposition between autonomy and vulnerability, we should look for a complementarity between both concepts. This proposal, like relational autonomy, takes the context into account as a priority.

While the concept of vulnerability has appeared primarily in research ethics and public health ethics, there are other interesting applications in the health field. For example, one can analyse how it may play out in some clinical situations with a population that has traditionally been considered vulnerable. Thus, certain elements of my approach can be enhanced and complemented by other analyses. For example, the practical model of layers of vulnerability incorporates protection but also respect for people and their autonomy, that is, helps them become empowered.[13] Incorporating a relational analysis of autonomy adds a new element to the traditional approach to vulnerability—that of providing protection—and implies a different strategy. This vision of vulnerability, non-existent in traditional approaches that revolved solely around protection (such as Goodin's (1986) pioneering approach), can be complemented and enhanced by a relational analysis of autonomy. It is precisely because the identification of vulnerabilities is the idea of protection that the opposition between autonomy and vulnerability is assumed. Thus, those who need to be protected are those who do not have full autonomy (hence the identification of persons who cannot consent, minors, persons in a subordinate relationship, etc.). These are people whom the traditional approach considers to have failures in the possibility of consenting or being autonomous. However, to remain in a vision that only considers protection may lead not only to paternalism but also to losing fundamental elements in the decision-making factor of people. Those of us who include autonomy and respect for persons, are betting on the possibility of expanding capacities and not closing them with a solely protective view with its paternalistic implications.

In an article co-authored with Paola Buedo we explore these notions and analyse the case of patients with mental distress and the decision making that can take place (Buedo and Luna 2021). We think of this group of people especially because the health system tends to underestimate the decision-making capacity of people with mental illness (Buedo and Luna 2021). A prevalent stereotype is to characterise these people as vulnerable, which implies a stigmatising attitude towards them. This led us to develop another type of approach linking relational autonomy with the layered concept of vulnerability. The article argues that labelling mentally ill individuals as 'vulnerable' reinforces stigma, perpetuating the flawed traditional view. The close association between mental illness and social stigma (Mascayano Tapia et al. 2015; Ben-Zeev et al. 2010; Corrigan et al. 2005; Castro 2005; Link and Phelan 2001) is shaped by different negative images, such as dangerousness, uselessness, impulsiveness and disability, among others, which undermine the decision-making capacity of the person with mental distress, their moral agency and the exercise of their autonomy.

It is here that the counterpoint between vulnerability and traditional autonomy lies. The traditional conception of autonomy, characterised as personal autonomy,

[13] The approach of Meek Lange et al. also supports the importance of empowerment, since the authors work closely with Catriona Mackenzie.

emphasises cognitive capacities and rationality, so mentally ill people are considered vulnerable and it is claimed that they are not autonomous, since their cognitive capacity is restricted. In this type of approach, all people with mental suffering are vulnerable, in an all-or-nothing logic: one either has or does not have autonomy, one is or is not vulnerable. On the other hand, relational autonomy and vulnerability, considered from the notion of layers, offer a very different perspective. They share fundamental points:

a) they are both dynamic and contextual,
b) they are reflected upon from the singularity of the subject, and.
c) they are not considered in a binary way (autonomy: yes or no, vulnerability: yes or no).

These three conceptual nodes of relational autonomy and vulnerability allow for the construction of a fertile conjunction of both (Buedo and Luna 2021: 1099). Including the non-ideal in the reconstruction of the concepts introduces new possibilities, such as promoting relational autonomy in general terms to respond to vulnerability in certain clinical situations. This would involve taking into account the context and its impact on people's decision making.

While this vision is relevant across all clinical disciplines, because patients do not make decisions in isolation but consult with their family and friends. Yet we also pointed out that it could be especially beneficial for people suffering from mental illness. One way to work with both concepts in a practical way is through the process of shared decision making (Slade 2017, Barani and Kopitowski 2013). Shared decision making is a form of collaboration between professionals, patients and family members. This process opens up the possibility for the patient to learn useful information and discuss it with their family and healthcare team, which in turn implies attributing to them a fundamental role in the construction of the diagnosis and treatment (Buedo and Luna 2021: 1089). This perspective allows to analyse which conditions could improve or damage the exercise of autonomy and to consider help and interaction with other loved or respected people in their environment who can collaborate positively in decision making. This in turn allows to think in a different way about the possibility of decision making in the clinical setting and to move away from the stigmatisation and infantilisation of people with mental illnesses. Thus, the person with mental suffering is not viewed in isolation, but with their social and emotional links, in order to work on possible clinical, reintegration or preventive treatments. A good interaction with the context can help to empower these people and consequently bring about improvements and some recovery.

In the article I have mentioned (Buedo and Luna 2021), we take a real case and analyse it from this perspective, showing not only the decision-making possibilities of the patient but also the possible strategies revealed by this analysis of vulnerability and the range of possible layers that can be attacked in different ways. We present the case of Sofia, 25 years old, diagnosed with an unspecified psychosis characterised by the presence of delusions or hallucinations. This type of condition generated significant social disability (Buedo and Luna 2021: 1093). Sofia had been doing well overall. She was in the final year of her biochemistry degree, participated in various activities, lived with and cared for her mother—who suffered from severe cognitive impairment—and had four brothers. When she experienced psychiatric episodes involving hallucinations, certain aspects of her social life came to a halt. This was the case with her studies and most of her social activities. It became evident that her first layer of vulnerability, which functioned as a cascade, was her inability to finish her university degree. This made it difficult for her to find employment, which in turn led to economic dependence and limited her opportunities to socialise, among other effects. Another layer consisted of the infantilising treatment she received from her older siblings, who did not allow her to engage in activities outside the home. These and other vulnerabilities can be identified and carefully worked on in order to avoid their actualisation or to minimise them (stimulus or trigger conditions could be distinguished).

The relational autonomy approach, however deals with empowerment and also with Sofia's positive relationships: she had very good relationships with her younger brother and others who can help her to overcome her low self-esteem and resume the life she enjoyed before the manifestation of her mental suffering. Thus, this strategy can be used to empower such patients and help in their treatment and improvement (Buedo and Luna 2021). Our analysis shows on a practical level how a model of vulnerability and relational autonomy can have a concrete impact in the psychiatric setting. Along the same lines, Buedo continued working together with Nair Della Vedova (Della Vedova, unpublished 2022) on her Master's thesis, broadening the scope of this proposal to a relational type of informed consent. This novel kind of approach enriches reflection in the clinical field.

8.8 Conclusion

In this chapter, I have taken up another fundamental concept that is the backbone of bioethics: autonomy. The chapter critiques the idealisations and limitations of the traditional conception of autonomy, but also shows how the concept was recovered and informed by the scrutinising gaze of feminism through the richness of multiple feminist proposals, from Susan Sherwin through Catriona Mackenzie to Marina Oshana. The rich analysis reveals the complex borders and dimensions of this concept, including how external situations of oppression or cultural situations affect it. Finally, looking at these problems from the South, we saw how a critical approach emerged from the periphery. It embodies the critique of the critique, looking at the body and linking it with the biases and oppressions introduced by others immersed in conservative and patriarchal societies.

Autonomy, reconceptualised within the non-ideal perspective, can be constructively related to the notion of vulnerability by functioning as a complement, rather than a counterpoint. This proposal is not only of theoretical interest but also bears practical fruit. Thus, this chapter ends showing the application of both layered vulnerability and relational autonomy.

References

Alvarado, M., ed. 2020. *Feminismos del Sur. Recorridos, itinerarios, junturas.* Prometeo.

Álvarez Medina, S. 2022. La ilusión de la autonomía plena. In *Autonomía y feminismos*, ed. M. Cavallo and A. Ramón Michel. Ediciones Didot.

Barani, M., and K. Kopitowski. 2013. Toma de decisiones compartidas: centrando los cuidados médicos realmente en nuestros pacientes. *Revista del Hospital Italiano de Buenos Aires* 33 (2): 60–64.

Beauchamp, T., and J. Childress. 1999. *Principios de ética biomédica.* Masson.

Belli, L., and D. Suárez Tomé. 2021. La autonomía revisitada desde la perspectiva de una bioética feminista. In *Tratado de género, derechos & justicia*, ed. M. Herrera et al. Rubinzal-Colzoni Editores.

Benson, F. 2005. Feminist intuitions and the normative substance of autonomy. In *Personal autonomy: New essays on personal autonomy and its role in contemporary moral philosophy*, ed. J. Stacey Taylor, 124–142. New York: Cambridge University Press.

Ben-Zeev, D., et al. 2010. DSM-V and the stigma of mental illness. *Journal of Mental Health* 19 (4): 318–327.

Bidaseca, K., and V. Vazquez Laba, eds. 2011. *Feminismos & Poscolonialidad. Descolonizando el Feminismo desde & el América Latina.* Ediciones Godot.

Buedo, P., and F. Luna. 2021. Toma de decisiones compartidas en salud mental: una propuesta novedosa. *Medicina & ética* 32 (4): 1087–1110.

Butler, J. 2009. *Dar cuenta de sí mismo.* Amorrortu: Violencia ética & responsabilidad.

Carter, S., et al. 2015. Relational conceptions of paternalism: A way to rebut nanny-state accusations and evaluate public health interventions. *Public Health* 129:1021–1029.

Castro, H. 2005. Estigma & enfermedad mental: un punto de vista histórico-social. *Revista de psiquiatría & salud mental Hermilio Valdizan* VI (1): 33–42.

Christman, J. 2004. Relational autonomy, liberal individualism, and the social constitution of selves. *Philosophical Studies: An International Journal for Philosophy in the Analytic Tradition* 117 (1/2): 143–164.

Ciriza, A. 2015. Construir genealogías feministas desde el Sur: encrucijadas & tensiones. *Revista MILLCAYAC* 2 (3): 83–104.

Corrigan, P., et al. 2005. The stigma of mental illness: Explanatory models and methods for change. *Applied and Preventive Psychology* 11 (3): 179–190.

Della Vedova, N. 2022. *Hacia un consentimiento informado relacional.* FLACSO: Tesis de Maestría en Bioética.

Dworkin, G. 1988. *The theory and practice of autonomy.* Cambridge University Press.

Espinosa Miñoso, Y. 2017. Hacia la construcción de la historia de un (des) encuentro: La razón feminista & la agencia antirracista & decolonial en Abya Yala. In *Praxis*, vol. 75. Revista de filosofía.

Fineman, M. 2008. The vulnerable subject: Anchoring equality in the human condition. *Yale Journal of Law and Feminism* 20 (1): 1–23.

Fineman, M. 2010. The vulnerable subject and the responsive state. *Emory Law Journal* 60:251–275.

Frankfurt, H. 1988. *The importance of what we care about.* Cambridge University Press.

Friedman, M. 2003. *Autonomy, gender, politics.* Oxford University Press.

Fundación Soberanía Sanitaria (FSS). 2019. *Salud feminista. Soberanía de los cuerpos, poder y organización.* Tinta Limón.

Gilligan, C. 1985. *La moral & la teoría. Psicología del desarrollo femenino.* Fondo de Cultura Económica.

Goodin, R. 1986. *Protecting the vulnerable: A re-analysis of our social responsibilities.* University of Chicago Press.

Kohen, B. 2005. Ciudadanía & ética del cuidado. In *Búsqueda de sentido para una nueva política*, ed. E. Carrió and D. Maffía. Paidos.

Lagarde, M. 1997. *Claves feministas para el poderío & la autonomía de las mujeres*. Puntos de encuentro.
Levine, C., et al. 2004. The limitations of 'vulnerability' as a protection for human research participants. *American Journal of Bioethics* 4 (3): 44–49.
Link, B., and J. Phelan. 2001. Conceptualizing Stigma. *Annual Review of Sociology* 27:363–385.
Luna, F. 1995. Paternalism and the argument of illiterate people. *Bioethics* 9 (3/4): July 1995:283–290.
Luna, F. 2015. Rubens, corsets and taxonomies: A response to meek Lange, Rogers and Dodds. *Bioética* 26 (6): 448–450.
Mackenzie, C. 2014. The importance of relational autonomy and capabilities for an ethics of vulnerability. In *Vulnerability: New essays in ethics and feminist philosophy*, ed. C. Mackenzie et al., 33–59. Oxford University Press.
Mackenzie, C. 2018. Feminist innovation in philosophy: Relational autonomy and social justice. *Women's Studies International Forum* 72:144–151.
Mackenzie, C. 2022. Tres dimensiones de autonomía: un análisis relacional. In *Autonomía y feminismos*, ed. M. Cavallo and A. Ramón Michel. Ediciones Didot.
Mascayano Tapia, F., et al. 2015. Estigma hacia los trastornos mentales: características e intervenciones. *Salud Mental* 38 (1): 53–58.
Meyers, D. 2000. Intersectional identity and the authentic self? Opposites attract! In *Relational autonomy. Feminist perspectives on autonomy, agency, and the social self*, ed. C. Mackenzie and N. Stoljar. Oxford University Press.
Morelli, S. 2022. *Reformulaciones del concepto de autonomía desde una perspectiva feminista*. Tesis de Maestría en Bioética, FLACSO.
Nedelsky, J. 1993. Reconceiving rights as relationship. *Review of Constitutional Studies* 1 (1):
Nelson, H. 1992. Against caring. *The Journal of Clinical Ethics* 3 (1): 8–15.
Oshana, M. 2022. ¿Es la autonomía social relacional un ideal plausible? In *Autonomía y feminismos*, ed. M. Cavallo and A. Ramón Michel. Ediciones Didot.
Ramón Michel, A., and M. Cavallo. 2022. Introducción. In *Autonomía y feminismos*, ed. M. Cavallo and A. Ramón Michel. Ediciones Didot.
Salles, A. 2008. Enfoques éticos alternativos. In *Bioética: nuevas reflexiones sobre debates clásicos*, ed. F. Luna and A. Salles. Fondo de Cultura Económica (FCE).
Sherwin, S. 1998. *The politics of women's health. Exploring agency and autonomy*. Temple University Press.
Sherwin, S. 2014. Feminismo & bioética. *Debate feminista* 49:45–69.
Slade, M. 2017. Implementing shared decision making in routine mental health care. *World Psychiatry* 16 (2): 146–153.
Stoljar, N. 2000. Autonomy and the feminist intuition. In *Relational autonomy. Feminist perspectives on autonomy, agency, and the social self*, ed. C. Mackenzie and N. Stoljar. Oxford University Press.
Stoljar, N. 2013. Feminist perspectives on autonomy, The Stanford Encyclopedia of Philosophy. https://plato.stanford.edu/archives/sum2013/entries/feminism-autonomy/
Stoljar, N., and C. Mackenzie. 2022. Relational autonomy in feminist bioethics. In *Routledge handbook of feminist bioethics*, ed. S. Carter et al. Routledge.
Valls-Llobet, C. 2008. Ética, salud & acto médico. Una perspectiva de género. In *Género & Bioética*, ed. M. Casado et al. Fontamara.
Westlund, A. 2022. Abnegación & responsabilidad propia ¿es la deferencia compatible con la autonomía? In *Autonomía y feminismos*, ed. M. Cavallo and A. Ramón Michel. Ediciones Didot.

Epilogue

The reflections presented in these pages, structured by theories and concepts that can be viewed from both ideal and non-ideal perspectives, should lead us to rethink bioethics. In what I believe is a novel approach to the field, I propose a meta-analysis of bioethics itself, how it works, and the kind of decision-making mechanisms it has on hand.

The book begins by presenting two cases that fall within the same field: reproductive rights. These are resolved in very different ways. While one appeals to the non-ideal, the other resorts to a rather ideal response. Framing the first chapter as a question is intended to prompt the reader to reflect on the markedly different types of responses that are often offered within this discipline. In my analysis, I identify redeemable and valuable elements in both ideal and non-ideal approaches, and I attempt the difficult task of not aligning exclusively with any one established tradition. There is no one theoretical framework—whether ideal theory (IT) or non-ideal theory (NIT)—that can reliably guide us through the diverse array of cases and policies that arise in the real world. The first two cases are intended to illustrate precisely this point.

We must take the winding and often steep path that seeks to hold both theoretical positions in productive tension. Nonetheless, choices cannot be made arbitrarily: it is essential to be able to justify clearly why one takes either an ideal or a non-ideal approach. A first step into the realm of the ideal and the non-ideal is to equip oneself with criteria to discern which path to follow. This enables theoretical coherence in decision-making. To that end, I have developed an instrument consisting of five criteria as an initial step. This tool is designed to help us choose between ITs and NITs. I refer to it as the "decision-making mechanism", and I first apply it to the two cases discussed in the opening chapter, which serve to introduce the reader to the philosophical landscape. I continue to employ this tool throughout the work.

In keeping with the proposal not to privilege a single theoretical stance, I argue in the analysis of the first case that we should adopt a NIT approach. This case examines the issue of conscientious objection in relation to legally permitted abortions in pre-2015 Argentina. Had an IT been applied in that context, it would likely have

F. Luna, *The Ideal and Non-Ideal in Bioethics*, The International Library of Bioethics 115, https://doi.org/10.1007/978-3-032-14594-9

harmed the most vulnerable—women who had been raped and lacked resources—resulting in their re-victimisation. The second case explores the conduct of initial clinical research involving pregnant women in countries where abortion in cases of foetal malformation—or what is sometimes termed 'eugenic abortion'—is not legally permitted and/or access to such procedures is unavailable. In contrast to the first case, I argue that this situation requires an ideal theory approach. Here, IT protects the most disadvantaged and allows such studies to be conducted with participants who can be adequately cared for. Together, these two cases illustrate the practical value of drawing on both NIT and IT to address real-world challenges. They also demonstrate that disregarding either theoretical domain—ideal or non-ideal—can limit our analytical potential. In the process of analysing these cases and presenting the decision-making mechanism, certain "clues" or preliminary insights emerge regarding the underlying theoretical assumptions at work.

These clues or assumptions are developed further in the second chapter, which is the most theoretical part of the book. Drawing on recent discussions in political philosophy, I explore the roles of ITs and NITs, analysing their relevance and how they interact with the real world. I highlight feasibility problems and show that both approaches contain valuable and recoverable elements. In fact, rather than representing a strict dichotomy, there exists a continuum of theories. One particularly important point—with implications for how we evaluate public policies and responses to specific cases—is the distinction between good and bad idealisations. The latter, for instance, involve biases or the disregard of certain realities, thereby perpetuating illicit privileges within particular groups. As such, the choices we make in practice are neither simple nor straightforward; they demand careful reflection and thorough assessment.

The analysis and exemplification of non-ideal practices and policies are resumed in Chapters 3 and 4 through the lens of the COVID-19 pandemic experience. Unfortunately, what transpired during the pandemic exemplifies the workings of our less-than-ideal world. At the beginning of Chapter 3, I take as a point of departure the Hobbesian scenario and the prisoner's dilemma as they unfolded during the pandemic. I explore the potential for adopting a form of moderate nationalism, such as the 'flu risk' criterion, and once again apply the decision-making mechanism to assess it. It is worth noting that this position is not only far from the ideal, but regrettably, also far from what actually occurred. Another example, examined in Chapter 4, is COVAX. This international initiative represented an interesting and promising first step, grounded in values such as solidarity and equity. It did indeed alter historical patterns of behaviour, and in that respect, it is to be welcomed. However, it also displayed shortcomings that cannot be ignored. The most serious failing was its inability to achieve an efficient and equitable distribution. COVAX would have required substantial improvements to the existing framework—even though these improvements were, in many cases, relatively straightforward to implement.

However, not everything in the realm of the ideal and the non-ideal needs to be bleak or negative. In Chapter 2, I also introduce the possibility of working with alternative concepts such as dynamic duties. These are oriented towards transforming society and involve using our political imagination to expand our capacity to realise fundamental principles. This compelling concept is revisited in Chapter 5, where I outline possible practices that could move us closer to an ideal state of affairs. An improved version of COVAX—or a 'COVAX Plus'—could not only mitigate some of the effects of the pandemic but, more importantly, help lay the groundwork for implementing dynamic duties. Chapter 5 explores the potential for operationalising these obligations and considers the temporary lifting of patents during pandemics as one such measure. In line with the logic of dynamic duties, there is also a pressing need for a robust international body capable of regulating and constraining the power of dominant countries and pharmaceutical companies, and ensuring that a collaborative and equitable ethos prevails. To that end, the chapter analyses the possibility of establishing a truly equitable international pandemic treaty, supported by a strong legal mandate. As the global mismanagement of the pandemic made clear, non-ideal responses failed to transcend the *status quo*. And it is precisely in moments such as those brought about by the pandemic that we must strive more urgently toward the ideal. In this regard, international civil society must remain vigilant and engaged, capable of acting as a genuine agent of change. Sound policies on patents and a fair, enforceable pandemic treaty are fundamental to such progress.

Beyond ideal and non-ideal theories and proposals, there are certain concepts, closely tied to non-ideal contexts, that must be analysed with care. One such concept is vulnerability. In this case, the aim is to avoid essentialist views—such as the traditional notion of the 'vulnerable population', which ultimately relies on inappropriate idealisations. Thus, the third part of the book turns to problems rooted in our real, everyday world. I explore alternative ways of analysing key bioethical concepts, deliberately moving away from conventional and traditional interpretations, and instead examining them through the lens of non-ideality. I focus first on the concept of vulnerability, and then on that of autonomy. The strength of the conception of vulnerability I develop lies in its refusal to simplify or idealise research participants, research settings, or public health challenges. It embraces the challenge of offering a nuanced, non-ideal response to a complex problem—fully in line with the overarching approach of the book, which invites us to confront complexity rather than shy away from it. The central claim of this section is that the way we conceptualise vulnerability profoundly shapes how we attempt to resolve it. My proposal addresses conceptual limitations while also offering significant practical contributions. The theoretical strategy I develop reveals the layered nature of vulnerability and its potential for harm, and immerses us firmly in the non-ideal world. Its practical application is set out in Chapter 7, where I provide concrete guidelines for addressing these layered forms of vulnerability.

Another key concept in bioethics explored through a non-ideal lens is autonomy. On one hand, feminist thought has shed light on the implications, commitments, and applications of relational autonomy, emphasising how the non-ideal nature of the world results in failures—both individual and collective—to exercise autonomy effectively. Various feminist contributions have exposed the idealised assumptions embedded in traditional accounts of autonomy, underscoring the importance of attending to context. The feminist perspective identifies bad idealisations (as they were presented in Chapter 2) and thus shifts the focus towards a less idealised conception. On the other hand, rethinking these concepts through a non-ideal framework requires the relationship between the two key concepts examined in this final part of the book to be reconsidered. Faced with the conventional opposition often drawn between vulnerability and autonomy, one challenge is to move beyond this impasse or counterpoint. I propose rethinking these concepts from more nuanced and sophisticated perspectives. I show how relational autonomy, rather than standing in contrast to layered vulnerability, can in fact complement and constructively interact with it.

Across these chapters, the book has progressively uncovered some of the ideal and non-ideal elements that shape our analyses and evaluations of reality. The theoretical foundations have been made explicit and illustrated with practical examples. The book began with the more complex theoretical proposals and concluded by engaging with key concepts that are especially relevant in the non-ideal domain. Their theoretical implications, practical applications, and mutual interrelations have also been carefully addressed.

Two goals structure this thinking. First and foremost, it aims to move beyond some of the simplifying concepts or principles that initially structured bioethics. It invites us to deliberate critically, to reason, to carefully consider cases and contexts, and to reject formulaic or easy solutions.

Secondly, it is built on the conviction that a solid theoretical analysis will lead us to better practice. How we conceive the ideal and the non-ideal will influence which perspective we use to approach and solve concrete cases. How the notion of vulnerability is conceptualised will influence our approach to supposedly vulnerable populations. It explains why, rather than referring to populations, different types of vulnerabilities should be identified, and that for example, different public policies or safeguards to address each vulnerability should be proposed. These analyses have their rightful place in the maturing of this discipline.

I should also acknowledge that these are preliminary and exploratory reflections that are far from exhausting all theoretical or meta-bioethical analyses. On the contrary, they aim to open up new ways of thinking more deeply about the responses, analyses, and concepts we frequently use in this discipline.

This book's proposal goes against the current trend—not only in certain bioethical analyses but also in communication and politics—which revels in extremes, dichotomies, and black-or-white messaging. Today, moderate reflection—the kind that discerns the many shades of gray around us or appeals to Aristotle's golden mean—seems to have fallen out of favour. In contrast to the aforementioned tendencies, this work seeks to provide responses better suited to grappling with complexity, avoiding theoretical gimmicks that attempt to 'order' reality but ultimately act as intellectual straightjackets—ill-fitting constraints that the overwhelming and intricate nature of reality refuses to accommodate.

Thus, this book reconsiders the very foundations of bioethics, proposing different analyses and conceptualisations that not only elucidate theory but can also show us how we should proceed to make decisions in this imperfect world.

Index

F. Luna, *The Ideal and Non-Ideal in Bioethics*, The International Library of Bioethics 115, https://doi.org/10.1007/978-3-032-14594-9

MIX
Papier aus verantwortungsvollen Quellen
Paper from responsible sources
FSC® C105338

If you have any concerns about our products,
you can contact us on
ProductSafety@springernature.com

In case Publisher is established outside the EU,
the EU authorized representative is:
Springer Nature Customer Service Center GmbH
Europaplatz 3, 69115 Heidelberg, Germany

Printed by Libri Plureos GmbH
in Hamburg, Germany